NEUROMETHODS ■ 37

Apoptosis Techniques and Protocols

Second Edition

NEUROMETHODS

Series Editors: **Alan A. Boulton** and **Glen B. Baker**

NEUROMETHODS ■ 37

Apoptosis Techniques and Protocols

SECOND EDITION

Edited by

Andréa C. LeBlanc

Bloomfield Center for Research in Aging,
Lady Davis Institute for Medical Research,
Jewish General Hospital and
Department of Neurology and Neurosurgery,
McGill University, Montreal, Quebec, Canada

HUMANA PRESS ✸ TOTOWA, NEW JERSEY

Preface to the Series

When the President of Humana Press first suggested that a series on methods in the neurosciences might be useful, one of us (AAB) was quite skeptical; only after discussions with GBB and some searching both of memory and library shelves did it seem that perhaps the publisher was right. Although some excellent methods books had recently appeared, notably in neuroanatomy, it was a fact that there was a dearth in this particular field, a fact attested to by the alacrity and enthusiasm with which most of the contributors to this series accepted our invitations and suggested additional topics and areas. After a somewhat hesitant start, essentially in the neurochemistry section, the series has grown and will encompass neurochemistry, neuropsychiatry, neurology, neuropathology, neurogenetics, neuroethology, molecular neurobiology, animal models of nervous disease, and no doubt many more "neuros." Although we have tried to include adequate methodological detail and in many cases detailed protocols, we have also tried to include wherever possible a short introductory review of the methods and/or related substances, comparisons with other methods, and the relationship of the substances being analyzed to neurological and psychiatric disorders. Recognizing our own limitations, we have invited a guest editor to join with us on most volumes in order to ensure complete coverage of the field. These editors will add their specialized knowledge and competencies. We anticipate that this series will fill a gap; we can only hope that it will be filled appropriately and with the right amount of expertise with respect to each method, substance or group of substances, and area treated.

Alan A. Boulton
Glen B. Baker

Preface

Neuronal loss that occurs in the central nervous system as a result of injury or neurodegenerative diseases is a devastating problem because of the lack of regenerative ability of the human body to functionally replace these neurons. Although there is promise of eventual replacement of these lost neurons through stem cell technology, inhibition of neuronal cell death in many instances would also provide a chance to reestablish their function. In many neurodegenerative diseases, however, there is no clear understanding of the cell death mechanism. Considerable debate focuses on whether these neurons undergo apoptosis, necrosis, or another yet undefined mode of cell death. The enigma is complicated by cell-type and by insult- and species-specific mechanisms of neuronal cell death. It is obvious that the identification and characterization of the mechanism of neuronal cell death can be resolved only through extensive research of the several types of neurons. Only when a fundamental understanding of the mechanisms of neuronal demise is achieved will we be in a position to identify the key pathways involved in human neurodegenerative diseases.

Since publication of the first edition of the *Apoptosis Techniques and Protocols* in 1997, the study of apoptotic mechanisms has boomed, and a number of key proteins involved in neuronal apoptosis have been identified. It has become clear that Bax and the family of caspases that is regulated through mitochondrial cytochrome-*c* release or through an extrinsic receptor-mediated pathway are key pro-apoptotic regulators of neuronal cell death. The first three chapters of this book present comprehensive technical approaches to study Bax (Hsu and Smaili), cytochrome-*c* (Ethell and Green), and caspases (Bounhar, Tounekti, and LeBlanc) in neurons. The following two chapters describe two methods, viral infections (Maguire-Zeiss, Bowers, and Federoff) and microinjections (Zhang and LeBlanc), to assess the importance of apoptotic proteins in cultures of primary neurons and in brain. Though key regulators of apoptosis have been uncovered, undoubtedly there are additional factors involved in neuronal apoptosis. Therefore, we must continue to search for proteins that may be responsible for neuronal loss in neurodegenerative diseases. DNA microarray assay is one of the current techniques used to identify differentially expressed genes in human disease (Eastman and Loring) and in transgenic mice models of neuro-degeneration (Tucker and Estus) . These chapters provide helpful insight into the design of an appropriate experimental protocol and in the interpretation of data from these microarrays. It is evident that as pro-apoptotic proteins are discovered, inhibitors of their functions in neuronal apoptosis must be sought. Berry and Ashe describe the role of differentially expressed G3PDH as an early marker of apoptosis and how one can isolate drugs against such pro-apoptotic molecules.

Though extensive research on apoptosis is performed and key mechanisms provided for many cell types, these do not necessarily apply to neurons. Indeed, we must consider the very specialized architecture of the neuron. Neurons are compartmentalized and polarized cell types. It is possible that specific pathways of apoptosis may be restricted to a specialized compartment of these cells and that activation of certain apoptotic mechanisms could result in cellular dysfunction prior to complete cell death. It has recently been proposed that the apoptotic process may occur in synapses. The technical and theoretical aspects of neuronal compartmentalization and the study of synaptosis are explained in the next two chapters, by Campenot and colleagues and by Cole and Gylys.

As we define the mechanisms of apoptosis that regulate cell death of neurons in cultures, it is essential to extend the studies to the brain tissue of individuals who have suffered from neuro-degenerative diseases. It is hoped that this will lead to the identi-fication of key processes that may eventually be controlled to prevent neuronal demise. The chapters by Roth on *in situ* detection of apoptosis and Smith and colleagues on the role of oxidative stress and apoptosis in Alzheimer's disease describe technical approaches associated with *in situ* detection of apoptosis.

The field of apoptosis research has grown exponentially in the past few years, and it would be impossible to describe each aspect of apoptosis as it applies to neurons. The importance of signal transduction pathways, transgenic animal models to study apoptosis, the other Bcl-2 family members, and the role of cell cycle gene expression in neuronal apoptosis has not been addressed in this book. These are equally important aspects of neuronal apoptosis.

Apoptosis Techniques and Protocols, Second Edition is intended to provide a handbook for the laboratory as well as a description of the limitations and advantages of the techniques proposed. I hope that it will be useful to both new and seasoned investigators who have an interest in unraveling the molecular mechanisms of neuronal cell death.

Andréa C. LeBlanc

Contents

Contributors

PAULA C. ASHE, PhD • *ALviva Biopharmaceuticals Inc., Saskatoon, Saskatchewan, Canada*

CRAIG S. ATWOOD, PhD • *Department of Chemistry, Case Western Reserve University, Cleveland, OH*

YOKO AZUMAYA, MSc • *Department of Cell Biology, University of Alberta, Edmonton, Alberta, Canada*

MARK D. BERRY, PhD • *ALviva Biopharmaceuticals Inc., Saskatoon, Saskatchewan, Canada*

YOUNES BOUNHAR, BSc • *Bloomfield Center for Reseach in Aging, Lady Davis Institute for Medical Research, Jewish General Hospital; Department Neurology and Neurosurgery, McGill University, Montreal, Quebec, Canada*

WILLIAM J. BOWERS, PhD • *Center for Aging and Developmental Biology and Department of Neurology, University of Rochester School of Medicine and Dentistry, Rochester, NY*

ROBERT B. CAMPENOT, PhD • *Department of Cell Biology, University of Alberta, Edmonton, Alberta, Canada*

GREG M. COLE, PhD • *Departments of Medicine and Neurology, University of California at Los Angeles, Los Angeles; VAGLAHS, North Hills, CA*

P. SCOTT EASTMAN, PhD • *Incyte Genomics, San Mateo, CA*

HUBERT ENG, PhD • *Department of Cell Biology, University of Alberta, Edmonton, Alberta, Canada*

STEVEN ESTUS, PhD • *Department of Physiology and Sanders-Brown Center on Aging, University of Kentucky, Lexington, KY*

DOUGLAS W. ETHELL, PhD • *Division of Cellular Immunology, La Jolla Institute for Allergy and Immunology, San Diego, CA*

HOWARD J. FEDEROFF, MD, PhD • *Center for Aging and Developmental Biology, Department of Neurology; Department of Microbiology and Immunology, University of Rochester School of Medicine and Dentistry, Rochester, NY*

DOUGLAS R. GREEN, PhD • *Division of Cellular Immunology, La Jolla Institute for Allergy and Immunology, San Diego, CA*

KAREN GYLYS, PhD • *Department of Nursing, University of California at Los Angeles, Los Angeles; VAGLAHS, North Hills, CA*

AYALA HOCHMAN, PhD • *Department of Biochemistry, George S. Wise Faculty of Life Sciences, Tel Aviv University, Tel Aviv, Israel*

YI-TE HSU, PhD • *Department of Biochemistry and Molecular Biology, Medical University of South Carolina, Charleston, SC*

BARBARA KARTEN, PhD • *Department of Medicine, University of Alberta, Edmonton, Alberta, Canada*

ANDRÉA C. LEBLANC, PhD • *Bloomfield Center for Reseach in Aging, Lady Davis Insitute for Medical Research, Jewish General Hospital and Department of Neurology and Neurosurgery, McGill University, Montreal, Quebec, Canada*

JEANNE F. LORING, PhD • *Incyte Genomics, Palo Alto, CA; currently Arcos BioScience, SanMateo, CA*

KAREN LUND, BSc • *Department of Cell Biology, University of Alberta, Edmonton, Alberta, Canada*

BRONWYN L. MACINNIS, BSc • *Department of Cell Biology, University of Alberta, Edmonton, Alberta, Canada*

KATHLEEN A. MAGUIRE-ZEISS, PhD • *Center for Aging and Developmental Biology and Department of Neurology, University of Rochester School of Medicine and Dentistry, Rochester, NY*

GRACE MARTIN • *Department of Cell Biology, University of Alberta, Edmonton, Alberta, Canada*

AKIHIKO NUNOMURA, MD, PhD • *Department of Psychiatry and Neurology, Asahikawa Medical College, Asahikawa, Japan.*

MARK E. OBRENOVICH, BSc • *Institute of Pathology, Case Western Reserve University, Cleveland, OH*

GEORGE PERRY, PhD • *Institute of Pathology, Case Western Reserve University, Cleveland, OH*

ARUN K. RAINA, BSc • *Institute of Pathology, Case Western Reserve University, Cleveland, OH*

KEVIN A. ROTH, MD, PhD • *Department of Pathology and Immunology, Washington University School of Medicine, St. Louis, MO*

CATHERINE A. ROTTKAMP, BSc • *Institute of Pathology, Case Western Reserve University, Cleveland, OH*

LAWRENCE M. SAYRE, PhD • *Department of Chemistry, Case Western Reserve University, Cleveland, OH*

SHUN SHIMOHAMA, MD, PhD • *Department of Neurology, Graduate School of Medicine, Kyoto University, Kyoto, Japan*

SORAYA SMAILI, PhD • *Department of Pharmacology, Federal University of São Paulo, São Paulo, Brazil*

MARK A. SMITH, PhD • *Institute of Pathology, Case Western Reserve University, Cleveland, OH*

ATSUSHI TAKEDA, MD, PhD • *Department of Neurology, Tohoku University School of Medicine, Sendai, Japan*

OMAR TOUNEKTI, PhD • *Bloomfield Center for Reseach in Aging, Lady Davis Insitute for Medical Research, Jewish General Hospital, Montreal, Quebec, Canada*

H. MICHAEL TUCKER, PhD • *Department of Physiology and Sanders-Brown Center on Aging, University of Kentucky, Lexington, KY*

DENNIS E. VANCE, PhD • *Department of Biochemistry, University of Alberta, Edmonton, Alberta, Canada*

JEAN E. VANCE, PhD • *Department of Medicine, University of Alberta, Edmonton, Alberta, Canada*

RUSSELL C. WATTS • *Department of Medicine, University of Alberta, Edmonton, Alberta, Canada*

YAN ZHANG, MSc • *Bloomfield Center for Reseach in Aging, Lady Davis Insitute for Medical Research, Jewish General Hospital; Department of Neurology and Neurosurgery, McGill University, Montreal, Quebec, Canada*

XIONGWEI ZHU, MSc • *Institute of Pathology, Case Western Reserve University, Cleveland, OH*

Molecular Characterization of the Proapoptotic Protein Bax

Yi-Te Hsu and Soraya Smaili

1. Introduction

Bax is a proapoptotic member of the Bcl-2 family. Members of this family can promote either cell survival, as in the case of Bcl-2 and Bcl-X_L, or cell death, as in the case of Bax and Bak. Bax was first identified as a Bcl-2 binding partner by immunoprecipitation (1). Subsequently it was shown that overexpression of Bax can accelerate cell death in response to various apoptosis stimuli (2). Physiologically, Bax plays an important role in neuronal development and spermatogenesis. Animals that are deficient in Bax have increased number of neurons and the males are known to be sterile (3,4). Under pathological conditions such as cerebral and cardiac ischemia, upregulation of Bax has been reported in the afflicted area of the tissues, implicating the participation of this protein in promoting neuronal and cardiomyocytic cell death (5–7). In certain cases of human colorectal cancer, mutations were found in the gene encoding Bax, suggesting that inactivation of Bax promotes tumorigenesis by enabling the tumor cells to be less susceptible to cell death (8).

Structurally, Bax, like other members of the Bcl-2 family, shares a common feature of having three conserved regions known as the BH (Bcl-2 homology) domains 1–3 (9,10) (see Fig. 1). These domains have been shown to be important for the apoptosis regulatory functions of these Bcl-2 family proteins. In addition, Bax and a number of Bcl-2 family members also possess a hydrophobic segment at their carboxyl terminal ends. For Bcl-2, this hydrophobic

From: *Neuromethods, Vol. 37: Apoptosis Techniques and Protocols, 2nd Ed.*
Edited by: A. C. LeBlanc @ Humana Press, Inc., Totowa, NJ

segment is required for anchoring the protein to various organelles, including endoplasmic reticulum, mitochondria, and nuclear outer membranes *(11,12)*. The three-dimensional structures of the Bax and its pro-survival antagonist Bcl-X$_L$ have recently been deciphered *(13,14)*. These two proteins appear to share a significant structural homology with the translocation domain of diphtheria toxin, especially at a helical loop domain formed by α-helices 5 and 6. This particular domain of diphtheria toxin has been shown to insert into lipid bilayer to form pores *(15–17)*.

In healthy cells, Bax is predominantly a soluble monomeric protein *(18–20)* despite the fact that it posseses a C-terminal hydrophobic segment. This hydrophobic domain, unlike those of Bcl-2 and Bcl-X$_L$, is sequestered inside a hydrophobic cleft *(13)*. Upon the induction of apoptosis by a variety of agents, a significant fraction of Bax was observed to translocate from the cytosol to the membrane fractions, in particular, the mitochondrial membrane *(18,20–28)*. This translocation process appears to involve a conformational change in Bax leading to the exposure of its C-terminal hydrophobic domain *(21,29)*. Deletion of the Bax C-terminal hydrophobic domain abrogated the ability of the mutant protein to translocate to mitochondria and greatly attenuated its ability to promote cell death *(21)*. On the other hand, point mutations of Bax that target the expressed proteins to mitochondria greatly increased Bax toxicity *(29)*. The translocation of Bax to mitochondria is associated with the release of cytochrome-*c* and the loss of mitochondrial membrane potential *(30–32)*. These phenomena may be related to the recent observations that Bax can form ion channels or pores in mitochondrial membranes *(33,34)* and that it can also modulate the activity of the mitochondrial permeability transition pores through binding to the pore components VDAC channel *(35)* or adenine nucleotide translocase *(36)*. Cytochrome-c activates caspase-3, leading to the proteolysis of the cell *(37)*, while the loss of mitochondrial membrane potential results in a decrease in cellular energy production. The proapoptotic activity of Bax, however, can be counteracted by co-expression with pro-survival factors Bcl-2 and Bcl-X$_L$, which can block Bax translocation to mitochondria during apoptosis *(24,38)*.

In this chapter, we describe the methodology we employ to study the translocation of Bax during apoptosis. Specifically, we will detail the techniques involved in the tracking of intracellular Bax distribution and Bax immunoaffinity purification.

2. Methods

2.1. Determination of Bax Intracellular Localization

The intracellular distribution of Bax in healthy cells and in cells undergoing apoptosis can be determined by subcellular fractionation, green fluorescent protein tagging, and immunofluorescence microscopy using anti-Bax antibodies.

2.1.1. Localization of Bax by Subcellular Fractionation

During apoptosis, a significant fraction of Bax has been shown to translocate from the cytosol to membranes, in particular mitochondrial membranes. This can be shown by subjecting healthy or apoptotic cells to hypotonic lysis, Dounce homogenization, and differential centrifugation. The resulting protein samples are then analyzed by Western blotting with anti-Bax antibodies.

1. Treat the cells of choice with the desired apoptosis inducer. Typically, we treat thymocytes with 1 µM dexamethasone and HL 60 promyelocytic leukemia cells with 1 µM staurosporine.
2. Collect the cells, pellet them by centrifugation, and wash them with the culture medium minus fetal bovine serum.
3. Resuspend the cells in the lysis buffer (10 mM HEPES, pH 7.4, 38 mM NaCl, 25-µg/mL phenylmethylsulfonyl fluoride, 1-µg/mL leupeptin, and 1-µg/mL aprotinin) at a cell density of 1–5 × 10^7 cells/mL. Incubate the mixture on ice for 15 min.
4. Homogenize the cells in a Dounce homogenizer until greater than 95% of the cells are broken. Centrifuge the lysate at 900g to pellet the nuclei in a Sorval SA-600 rotor. Save the pellet.
5. Centrifuge the postnuclear supernatant at 130,000g in a Beckman Ti 80 rotor to pellet the membranes.
6. Resuspend both the nuclear and crude membrane pellets with lysis buffer equal to the volume of the supernatant.
7. Analyze the protein samples by a 12% sodium dodecyl sulfate (SDS) polyacrylamide gel, followed by Western blotting analysis with an anti-Bax antibody. Typically, we transfer the protein samples from the SDS gel onto Immobilon-P membranes (Millipore) in Tris-glycine buffer containing 10% methanol. The membranes are blocked in the blocking buffer (phosphate-buffered saline [PBS] containing 5% fetal bovine serum and 0.05% Tween-20) for 90 min. The blots are then incubated in the primary antibody (1/10 dilution in the blocking buffer) for 45 min,

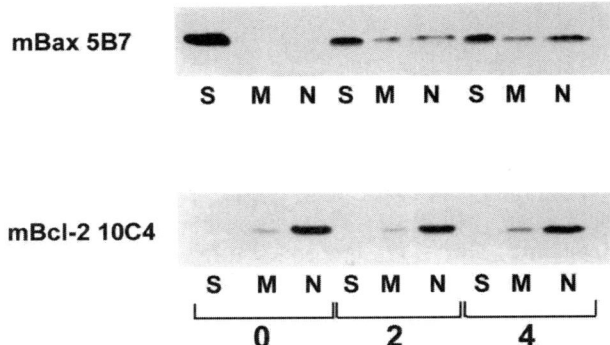

Fig. 1. Bax translocates from the cytosol to membranes during apoptosis. *Methods:* Apoptosis was induced in murine thymocytes by the addition of 2 µM dexamethasone. Untreated cells and cells incubated for 2 h and 4 h with dexamethasone were hypotonically lyzed, homogenized with a Dounce homogenizer, and separated into soluble protein (S), high-speed membrane pellet (M), and nuclear fractions (N) by differential centrifugation. The protein samples were analyzed by Western blotting with anti-murine Bax 5B7 and anti-murine Bcl-2 10C4 monoclonal antibodies.

washed in 0.05% Tween/PBS, and further incubated with the horseradish peroxidase-conjugated sheep anti-mouse Ig secondary antibody (Amersham Pharmacia Biotech) for 30 min. The blots are washed with 0.05% Tween/PBS and then with PBS and developed by the ECL chemiluminescent method (Amersham Pharmacia Biotech). Figure 1 is an example of the subcellular fractionation of murine thymocytes treated for 2 and 4 h with 1 µM dexamethasone.

2.1.2. Tracking Bax Movement by GFP Tagging

The tagging of green fluorescent protein (GFP) to Bax allows us to monitor the intracellular redistribution of Bax during apoptosis. To carry out this study, we subcloned the gene encoding human Bax into the 3' end of a GFP expression vector. This GFP/Bax fusion construct can then be transfected into mammalian cells for visualization by confocal or fluorescence microscopy. By staining the transfected cells with Mitotracker Red, tetramethylrhodamine ethyl ester (TMRE), or *bis*-benzamide, one can simultaneously track Bax localization with respect to mitochondrial distribution, mitochondrial membrane potential, and nuclear morphology.

2.1.2.1. GENERATION OF GFP-BAX CONSTRUCTS

1. PCR human Bax from human Bax in BS plasmid (gift of Dr. Stanley Korsmeyer) with flanking primers containing the appropriate restriction sites for subcloning into the C3-EGFP vector (Clontech). Typically, we insert the PCR fragment between the *Hind III* and *Eco RI* sites in the polycloning regions of the vector. To make Bax mutants, one can apply a two-step polymerase chain reaction (PCR) strategy by using two internal primers containing the desired mutation sites. Digest both the PCR fragment and the vector with restriction enzymes and then ligate them together using the Rapid DNA Ligation Kit (Boehringer Mannheim).
2. Transform *Escherichia coli* with the ligated products and plate out the cells onto LB agar plates containing 30 µg/mL kanamycin. Pick 12 colonies, grow them up overnight in LB containing 30 µg/mL kanamycin, and do a miniprep on each culture using the Wizard Plus Minipreps DNA purification system (Promega). Digest 8 µL of the plasmid preparations with the restriction enzymes and analyze the digested samples over a 1.5% agarose gel.
3. For a clone that contains the correct insert (as determined by restriction digest and DNA sequencing), prepare a 200-mL culture and carry out a large-scale plasmid DNA preparation using the Qiafilter Maxiprep kit (Qiagen). Resuspend the final DNA pellet in 400 µL sterile Tris ethylenediamine tetraacetic acid (TE) buffer, take a small aliquot (10 µL diluted in 490 µL TE) and determine DNA concentration by measuring the A_{260} of the sample. Typically, for the plasmid DNA, an A_{260} reading of 1 equals 50 µg/mL DNA. Store plasmid DNA samples at −70°C.

2.1.2.2. TRANSFECTION AND VISUALIZATION OF GFP-BAX CONSTRUCTS

1. Cells to be transfected are grown in 2-well chamber slides (Nalgene) pretreated with 5-µg/mL poly-L-lysine. Typically, we use L929 and Cos-7 green monkey kidney epithelial cells (ATCC) for our studies and they are maintained in minimum essential and Dulbecco's modified Eagle media (Life Technologies), respectively. The cells are cultured at 37°C in 5% CO_2 and split twice weekly using 0.05% trypsin/0.02% versene (Biofluids)
2. Transfect the cells with 0.5 µg C3-GFP-Bax plasmid DNA per well with lipid reagents using the protocols described by the

manufacturers. We have tested a number of transfection reagents, including LipofectAmine (Invitrogen), FuGENE (Roche), and Superfect (Qiagen) and found them to be all suitable for transfecting GFP-Bax into Cos-7 and L929 cells. For co-transfection studies to determine how expression of other proteins (i.e., Bcl-2 and Bcl-X_L) affect Bax translocation, we generally use 0.5 µg GFP-Bax plasmid with 1.5 µg of the plasmid encoding the protein of interest. It may be necessary first to determine the optimal amount of lipid reagents to use in different cell types. After transfection, incubate the cells overnight at 37°C in 5% CO_2.

3. Prepare the cells for microscopy by washing the cells one time with the growth medium and stain the cells with 20 ng/mL mitochondrial-specific dye Mitotracker Red CMXRos (Molecular Probes) and/or 100 ng/mL *bis*-benzamide for 30 min. The cells are then subjected to treatment with an apoptosis inducer. We commonly use 0.5–1 µM kinase inhibitor staurosporine to induce apoptosis in Cos-7 and L929 cells.

4. Visualize the cells over a fluorescence or confocal microscope (Fig. 2). In our studies, we use a Zeiss LSM 410 microscope with a 40 × 12 NA Apochromat objective. The 488- and 568-nm lines of a krypton/argon laser are used for excitation of GFP and Mitotracker, respectively. The 364-nm line of an argon laser is used for excitation of *bis*-benzamide.

2.1.2.3. Simultaneous Tracking of Bax Localization and Mitochondrial Membrane Potential

1. For simultaneous measurements of Bax localization and movement and mitochondrial membrane potential ($\Delta\Psi_m$), cells are plated on polylysine-coated cover slips. These cover slips are removed from culture dishes right before the experiment and

---

Fig. 2. Mitochondria are the primary sites for Bax insertion during apoptosis. (**A**) Bax translocation from the cytosol to mitochondria is an early event in apoptosis. *Methods:* Cos-7 cells transfected with GFP-Bax were treated with 1 µM staurosporine and tracked over time by confocal microscopy. The two GFP-Bax expressing Cos-7 cells in the observation field initially displayed a green diffuse pattern indicative of its cytosolic localization. Upon apoptosis induction by staurosporine, GFP-Bax shifts from a diffuse cytosolic pattern to a punctate pattern (panels *a*–*e*). The Cos-7 cells were also stained with 100 ng/mL of the nuclear stain *bis*-benzamide

A

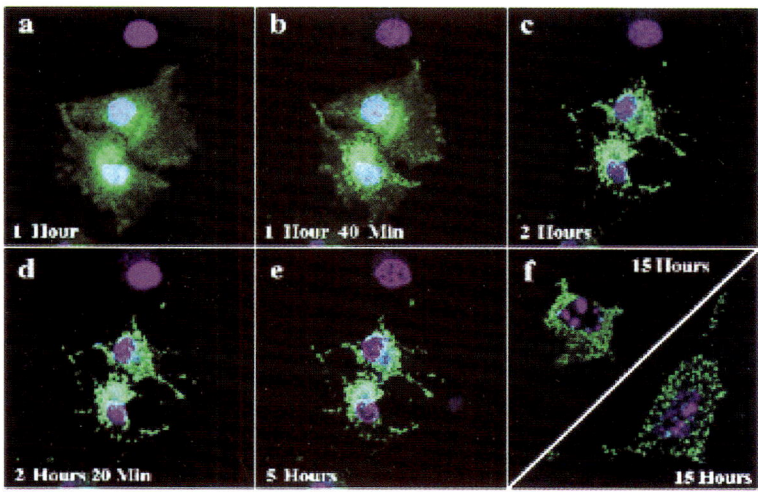

B

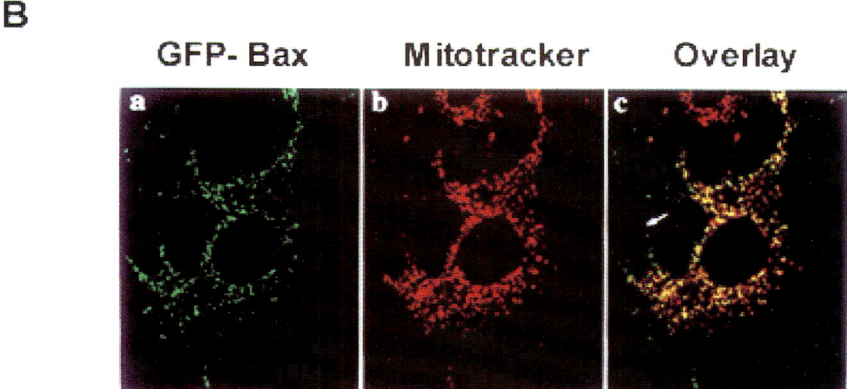

(shown in blue). Nuclear fragmentation associated with apoptosis was not observed until a much later time after GFP-Bax insertion (panel f). **(B)** Membrane-bound GFP-Bax co-localizes with mitochondria during apoptosis. *Methods:* Cos-7 cells transfected with GFP-Bax were treated with $1 \mu M$ staurosporine to induce apoptosis. The cells were stained with 20 ng/mL Mitotracker Red CMXRos and examined 4 h later by laser confocal microscopy at the appropriate wavelengths for GFP (*a*) and Mitotracker (*b*). The two images were then overlaid (*c*) to show that a majority of the punctate GFP-Bax is localized to the mitochondria.

placed in a Leiden chamber and washed with the microscopy buffer (130 mM NaCl, 5.36 mM KCl, 0.8 mM MgCl$_2$, 1.5 mM CaCl$_2$, 1 mM Na$_2$HPO$_4$, 2.5 mM NaHCO$_3$, 25 mM glucose, 20 mM HEPES, 1 mM Na-pyruvate, 1 mM sodium ascorbate, pH 7.3) before visualization.

2. For membrane potential change ($\Delta\Psi_m$) measurements, cells are loaded with 25–50 nmol of TMRE (Molecular Probes), which is a cationic potentiometric dye that accumulates preferentially in energized mitochondria driven by $\Delta\Psi_m$. TMRE is rapidly and reversibly taken up by live cells, and provides a quantitative and qualitative measure of $\Delta\Psi_m$ changes.

3. Acquire fluorescence images using a microchannel plate intensifier with a CCD camera (Fig. 3). For simultaneous measurements of GFP-Bax movement and $\Delta\Psi_m$ with TMRE, images are acquired using a filter set fitted with a dual-band dichroic mirror (520 and 575 nm) and bandpass filters (520 and 600 nm, 40 nm full width at half-maximum transmission; Chroma Technologies). Cells are excited through alternating narrow-bandpass filters with excitation wavelengths for GFP (485 nm) and TMRE (530 nm). Images are acquired every 120 s, with the image intensifier gain being independently adjusted for the two fluorophores by computer control.

2.1.3. Visualization of Bax
by Immuofluorescence Microscopy

With the availability of high-quality anti-Bax antibodies, it is possible to monitor Bax localization in a Bax overexpression system and in cell lines that express relatively high levels of endogenous Bax. We found that the N-terminal end of Bax appears to be the most readily accessible site for antibody binding across different species. For human Bax, two anti-Bax antibodies 1F6 (a.a. 3–16) and 6A7 (a.a. 12–24) have been used successfully for the immunohistochemical labeling of Bax *(29)*.

1. Treat the cells grown in chamber slides with or without the apoptosis inducer for designated time periods.
2. Wash the cells 3X with PBS and fix them in freshly prepared 3% paraformaldehyde in PBS for 10 min.
3. Wash the fixed cells again 3X with PBS and permeabilize the cells in 0.15% Triton X-100 or 0.04% saponin in PBS for 15 min.

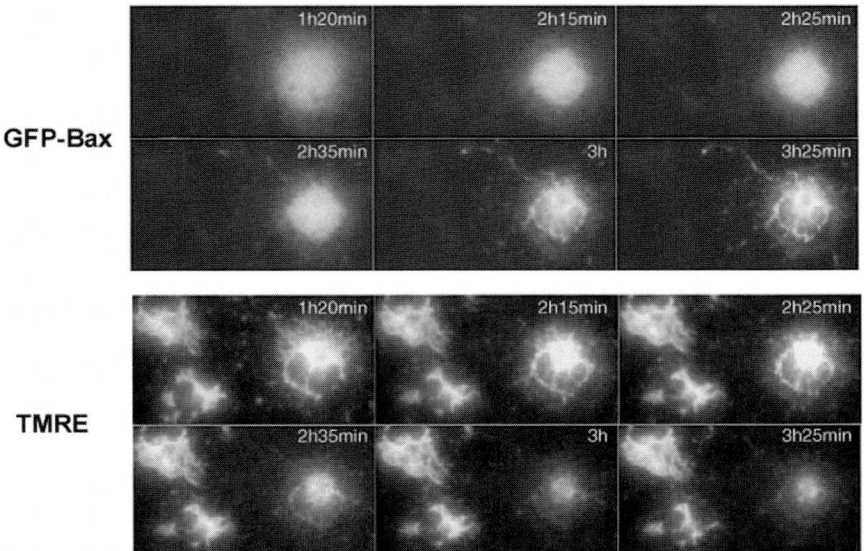

Fig. 3. The loss of mitochondrial membrane potential is associated with Bax translocation to mitochondria. *Methods:* Healthy Cos-7 cells expressing GFP-Bax were stained with TMRE for mitochondrial membrane potential visualization. These cells were treated with 1 μM staurosporine and the fluorescence patterns of GFP-Bax (rows 1 and 2) and TMRE (rows 3 and 4) were simultaneously monitored over time (shown on the upper right-hand corner) with a CCD camera. For this particular experiment, the cell on the right is transfected with GFP-Bax while the two cells on the left represent untransfected cells. The two untransfected cells are not visible in the top two rows showing GFP-Bax fluorescence but can be seen in the lower two rows for TMRE staining. As shown in the upper two rows, GFP-Bax shifted from a diffuse cytosolic state to a punctate mitochondrial-bound localization upon the induction of apoptosis by staurosporine. In terms of mitochondrial membrane potential change during apoptosis, the presence of GFP-Bax promoted the loss of mitochondrial membrane potential in the transfected cell as indicated by a decrease in the TMRE fluorescence over time (rows 3 and 4). This loss in mitochondrial membrane potential appeared to have occurred concomitantly with GFP-Bax translocation to mitochondria. There was very little loss of TMRE fluorescence in the two untransfected cells on the left within the time frame examined.

Table 1
Generation of Anti-Bax Monoclonal Antibodies

Antibody	Epitope (amino acid)	Species crossreactivity
mBax 5B7	3–16	Mouse
uBax 6A7	12–24	Mouse, human, rat, cow, and monkey
uBax 2C8	43–62	Mouse, human, rat, cow, and monkey
rBax 1D1	3–16	Rat
hBax 1F6/2D2	3–16	Human, bovine, monkey
hBax 1H2	43–62	Human, bovine, monkey

Methods. Anti-Bax monoclonal antibodies were generated by immunizing mice with synthetic peptides corresponding to different regions and species of Bax. Splenocytes taken from immunoreactive mice were then fused with NS-1 myeloma cells. The resulting hybridomas were then screened by ELISA with glutathione S-transferase (GST)-Bax fusion proteins and the species crossreactivities of the antibodies were determined by Western blotting.

4. Block the cells for 30 min in the blocking buffer (10% fetal bovine serum in PBS).
5. Incubate the cells with 1 mL anti-Bax antibodies 1F6 or 6A7 (1/2 dilution in 2% fetal bovine serum) for 1 h.
6. Wash the cells 3X with PBS at 10-min intervals and incubate the cells in rhodamine-labeled goat anti-mouse IgG (1/200 dilution in PBS/5% fetal bovine serum). Wash the cells 3X at 10-min intervals. Mount the slides with SlowFade antifade kit (Molecular Probes) and visualize the stained cells by fluorescence or confocal microscopy.

2.2. Immunoprecipitation of Bax

Since Bax is present in relatively low quantities inside the cell, affinity purification represents an important means for the biochemical characterization of this protein. In our laboratory, we routinely carry out the immunoaffinity purification of Bax using monoclonal antibodies. To date, we have generated six different anti-Bax monoclonal antibodies. These antibodies were raised against synthetic peptides corresponding to different regions and species of Bax (*see* Table 1). The use of synthetic peptides as immunogens enables us to predetermine the epitopes of these antibodies. Four of the antibodies that we have raised are directed against the N-terminal regions of human, mouse, and rat Bax, and they are the primary antibodies that we use for immunoprecipitation. Their epitopes, species specificity, and known applications are shown in Table 1. Immunoprecipitation of Bax involves the purification of anti-Bax antibodies from ascites fluids, conjugation of the antibody to Sepharose beads, incubation of the antibody beads with the cell lysate, and elution of the bound proteins.

2.2.1. Generation of Anti-Bax Monoclonal Antibodies

1. Design appropriate peptides of 12–18 amino acids in length for immunization. Generally we choose regions that are relative hydrophilic. All peptides we have synthesized have a cysteine residue at the N-terminal end to enable their conjugation to a carrier protein, in this case, keyhole limpet hemocyanin. We have synthesized six different peptides corresponding to different regions and species of Bax (*see* Table 1).
2. Mix 3 mg peptides (in 300 µL PBS) with 300 µL maleimide activated keyhole limpet hemocyanin (10 mg/mL; Pierce) and incubate at room temperature for 2 h. Load 600 µL of the conjugated product onto a PD-10 column that has been pre-equilibrated with PBS. This step removes unconjugated peptides and EDTA from the mixture. Continue to wash the column with PBS and collect 0.6-mL fractions. Generally, we collect 10 fractions and pool fractions 5–9. The pooled samples are aliquoted into 0.4-mL portions and stored at –20°C.
3. Immunize Balb/c mice intraperitoneally or subcutaneously every 3 wk with the peptide/KLH conjugate. Mix the immunogens 1/1 with Ribi's adjuvant (Corixa). Inject approximately 100 µL of the mixture per mouse (4 mice per peptide sample).
4. After the third or fourth injection, tail-bleed the mice and collect blood with heparin-coated capillary tubings. Let blood samples sit at room temperature for 1 h. Spin down the blood cells and collect the sera.
5. Assay the sera by Western blotting with cell lysate containing Bax. For the immunoreactive mouse, wait 1 mo, and do a final boost with 100 µL peptide conjugate, either intraperitoneally or intravenously.
6. Five days after the final boost, sacrifice the mouse, remove the spleen, and prepare splenocytes. Wash the splenocytes 3X with Iscove's medium to remove the serum. Count the cells. Generally, one gets $2–3 \times 10^8$ cells per spleen. Resuspend the splenocyte pellet in 10 mL Iscove's medium.

7. Collect NS-1 myeloma cells by centrifugation and count the number of NS-1 cells required for fusion. Generally, one needs 7–10 spleen cells per NS-1 cell.
8. Wash the NS-1 cells 3X with Iscove's medium to remove the serum and resuspend the final pellet in 10 mL Iscove's medium.
9. Pool the splenocytes with NS-1 cells and spin them down. Remove most of the liquid but leave about 100 μL behind. Resuspend the cell pellet by gentle tapping and fuse the cells with PEG (1500 daltons). Add 1 mL polyethylene glycol (PEG) (37°C) to the cell suspension over 30 s. Rotate the sample for 30 s and let it sit for another 30 s. Add 1 mL Iscove's medium over 60 s and another 4 mL over 30 s. Let the mixture sit for 2–3 min.
10. Spin down the cells and resuspend the pellet in 150 mL Iscove's medium containing 20% fetal bovine serum and 1X hypoxanthine, aminopterine, and thymidine (HAT).
11. Pipet 100 μL cell suspension per well into 96-well plates (about 15 plates) using a multichannel pipetter. On the third day, add another 100 μL of the growth medium into each well.
12. Screen for positive hybridoma cells on d 10 by ELISA. The ELISA plates are prepared by coating the microtiter plates (Falcon) with GST/Bax fusion proteins containing the specific peptide sequence used to immunize the mice. ELISA is carried out by blocking the wells with 2% fetal bovine serum in PBS for 1 h. The plates are then incubated with 100 μL culture supernatant (1/3 dilution in the blocking buffer) for 1 h and washed 4X with PBS. Following a 1-h incubation with 100-μL/well horseradish peroxidase-conjugated sheep anti-mouse Ig secondary antibody (1/1000 dilution; Amersham), the plates are washed and developed by the addition of ABTS peroxidase substrate solution (100 μL/well; Kirkegaard & Perry Laboratories). The positive wells turn dark green in color.
13. Further verify the ELISA positive clones by Western blotting analyses of the ELISA-positive culture fluids with Immobilon membrane strips containing Bax.
14. Subclone positive hybridoma cells in 96-well plates (0.5–2.5 cells/well) in the presence of spleen feeder cells. After 10 d, rescreen the clones by ELISA and Western blotting.
15. Select cells from positive wells containing single hybridoma colonies and expand these clones gradually. Once 6-well plate stage is reached, replace HAT with hypoxanthine and thymidine (HT) and amplify the cells in 10-cm plates.

16. Freeze down the amplified hybridoma cells in liquid nitrogen once sufficient cells have been grown. Culture supernatant from these cells can be used for Western blotting and immuno-fluorescence labeling studies.
17. Generate ascites fluid by priming mice with 0.5 mL pristane and injecting 2×10^6 hybridoma cells a week later. Collect ascites fluids in 10–14 d.

2.2.2. Preparation of Anti-Bax Monoclonal Antibody Beads

2.2.2.1. PURIFICATION OF MONOCLONAL ANTIBODIES FROM ASCITES FLUID

1. Remove precipitates from ascites fluid by centrifugation at 8000 rpm in a Sorvall SA 600 rotor for 10 min at 4°C.
2. Dilute ascites fluid 1/2 with PBS and mix it with equal volume of ice-cold saturated ammonium persulfate. Stir on ice for 10 min and let it sit for an additional 10 min on ice.
3. Pellet the antibody by spinning the sample at 8000 rpm for 15 min in a Sorvall SA 600 rotor.
4. Resuspend the pellet in 10 mM Tris, pH 7.9 (same volume as the PBS-diluted ascites fluid) and dialyze against 4 changes of 10 mM Tris, pH 7.9 (2 L each, and change the buffer solution every 4 h).
5. After dialysis, spin down the precipitated material by centrifugation at 8000 rpm for 10 min using a Sorvall SA 600 rotor. Load the supernatant onto a DEAE Sephacel anion-exchange column (Amersham Pharmacia Biotech) equilibrated with 10 mM Tris, pH 7.9. Typically we use 2 mL DEAE beads per milliliter of ascites fluid.
6. Wash the column extensively with the equilibration buffer and elute the antibodies off the column with a 0–0.3 M NaCl gradient (15 column volumes per buffer chamber).
7. Collect 1–2-mL fractions and determine the A_{280} of the antibody solution. Antibodies should be eluted within the first major A_{280} peak.
8. Analyze the peak fractions by a 10% SDS polyacrylamide gel. The samples should be boiled in a loading gel buffer containing β-mercaptoethanol. Stain the gel with Coomassie blue to visualize the antibody bands. Generally, one should see two major bands with molecular weights of 50 and 25 kDa, corresponding to the heavy and light chains of the antibody. Pool the peak antibody fractions.

2.2.2.2. COUPLING OF BAX ANTIBODIES TO SEPHAROSE BEADS

1. Dialyze the purified anti-Bax monoclonal antibodies against 2 L of 10 mM borate, pH 8.4, at 4°C. Change the dialyzing solution every 6–8 h for a total of 4 times. Remove the precipitate by centrifugation at 8000 rpm in a Sorvall SA 600 rotor. Determine the A_{280} of the sample. Typically, an A_{280} of 1.3 equals 1 mg/mL for antibodies.
2. Weigh out appropriate amount of CNBr-activated Sepharose 4 B beads (Sigma). Generally, we couple 1–2 mg of antibody per milliliter of beads. Soak the beads with the borate buffer for 30 min. Wash the beads 3X with the borate buffer and then mix the beads with the antibody solution. Gently rotate the mixture overnight at 4°C.
3. Spin the beads down and remove the supernatant. Check the A_{280} of the supernatant. Typically, the coupling efficiency is >95%. Block the unreacted sites on the beads by incubation with 20 mM Tris, pH 8.4, 50 mM glycine, and 100 mM NaCl overnight. Spin down the beads and wash them 3X with 20 mM Tris, pH 7.4. The beads can now be used for immunoprecipitation.

2.2.3. Immunoprecipitation of Bax
from Detergent-Solubilized Membranes

1. Solubilize the cells of choice in 10 mM HEPES, pH 7.4, 150 mM NaCl, 25 µg/mL phenylmethylsulfonyl fluoride (PMSF), and 1% detergent of choice (i.e., Triton X-100, Nonidet P-40, and Chaps) at a concentration of 1–5×10^7 cells/mL.
2. Spin down the unsolubilized material at 12,000 rpm using a Sorval SA 600 rotor.
3. Incubate 1–3 mL cell lysate with 150 µL antibody beads that have been pre-equilibrated in the solubilization buffer. Allow the incubation to proceed for 3 h at 4°C with constant but gentle rotations.
4. Spin the beads down, remove the supernatant, and wash the beads extensively with 2 washes of 10 mL solubilization buffer. Transfer the beads to a centrifugal filter unit (Millipore) and remove the remaining solution by brief centrifugation in a microfuge.
5. Elute Bax off the antibody beads with 180 µL of 0.1 M acetic acid containing 0.3% detergent of choice and neutralize the acid with 30 µL of 1 M Tris, pH 8.4.

2.2.4. Immunoaffinity Purification of Bax from the Cytosol

1. Prepare soluble protein extracts of murine thymocytes (1×10^8/mL in 10 mM HEPES, pH 7.4, 38 mM NaCl, 25 μg/mL phenylmethylsulfonyl fluoride, 1-μg/mL leupeptin, and 1-μg/mL aprotinin) by hypotonic lysis, Dounce homogenization, and differential centrifugation as described under Subheading 2.1.1. For the purification of human or bovine Bax, it is necessary to exclude NaCl.

2. Load the lysate (180 mL) onto a 20-mL AF-heparin-650M column equilibrated in the lysis buffer and collect the flow-through that contains Bax. This step removes a significant amount of soluble proteins from the extract.

3. Load the flow-through onto a 15-mL Fractogel EMD TMAE-650M anion exchange column equilibrated in the lysis buffer. Wash the column with 3 column volumes of the equilibration buffer. This step concentrates Bax for immunoaffinity chromatography while removing additional non–Bax soluble proteins from the extract.

4. Elute the bound protein sample which contains Bax with 3 column volumes of the elution buffer (lysis buffer with 125 mM NaCl).

5. Incubate the TMAE column eluant with 0.5 mL of anti-murine Bax 5B7 monoclonal antibody beads for 3 h at 4°C.

6. Remove the unbound proteins from the beads and wash the beads extensively with 50 column volumes of the wash buffer containing 150 mM NaCl.

7. Elute the bound Bax off the beads with 0.1 M acetic acid containing 0.1% Triton X-100. Collect 0.3-mL fractions and neutralize with 40 μL of 1 M Tris, pH 8.4. Triton X-100 is needed for elution because without the presence of a small quantity of detergent, the acid-exposed Bax tends to stick to Sepharose beads. An alternative elution method is to incubate the beads with 0.1 mg/mL synthetic peptides corresponding to the Bax antibody epitope. This elution method will likely yield native Bax, but is more expensive. The Bax purification procedure described here can produce relatively pure Bax preparations as shown in Fig. 4.

3. Discussion

In this chapter, we have described some commonly used protocols that we employ in our laboratory to characterize the proapoptotic protein Bax using biochemical, immunological, and cell biological approaches. Our goal is to understand the molecular mechanism by which Bax promotes cell death. The translocation of Bax from the cytosol to mitochondria represents an important mechanism by which Bax promotes cell death. Under Subheading 2.1., we described three different methodologies to examine Bax translocation during apoptosis. Each of these methods has its own advantages and disadvantages. The subcellular fractionation approach is a simple method to detect Bax translocation. However, it cannot determine the percentage of cells in which Bax translocation has occurred. In some cases, it may be difficult to detect minor Bax translocation if the Western blotting techniques are not performed optimally. Tagging GFP to Bax also represents an extremely simple technique to examine Bax translocation. This method has the advantage of allowing simultaneous monitoring of mitochondrial and nuclear morphologies and mitochondrial membrane potential. This approach, however, represents an overexpression system. Depending on cell types, overexpression of GFP/Bax may cause cell death. Thus, it is necessary first to optimize the transfection condition to minimize cell toxicity. Finally, immunofluorescence labeling with specific anti-Bax monoclonal antibodies appears to be an excellent method to examine Bax translocation. However, in certain cell types that express low levels of endogenous Bax, it may be difficult to detect Bax since immunofluorescence labeling is a less sensitive method compared to Western blotting.

The finding that Bax localization to mitochondria is essential for its function suggests that it may mediate apoptosis through specific mitochondrial receptors. Because Bax is present in low quantities inside the cell, the challenge lies in the isolation of its binding partner. To carry out this study correctly, it would be necessary to perform a large-scale immunoprecipitation with anti-Bax antibodies and analyze the immunoprecipitated products by Coomassie blue staining of the SDS polyacrylamide gels. Subsequently, one would need to carry out a reverse immunoprecipitation with antibodies against the Bax-binding protein and examine the association of these two proteins by Coomassie blue staining of the SDS polyacrylamide gel and Western blotting. The detection of Bax and its associated proteins following immunoprecipitation should always involve Coomassie blue staining. If the detection of Bax-binding protein is carried out by Western blotting only, one may easily miss other Bax-binding protein. Also, another major difficulty associated with identifying Bax-binding proteins is the fact

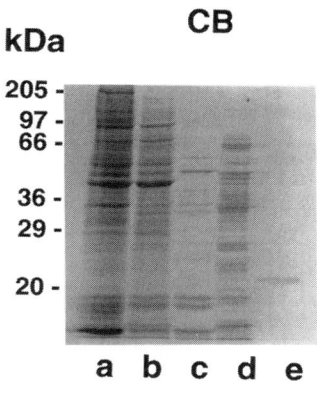

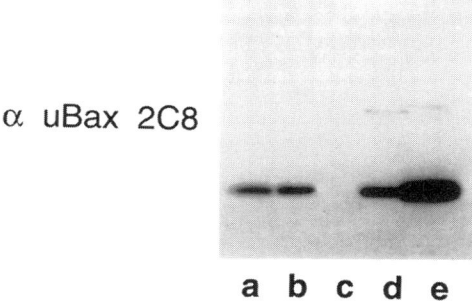

Fig. 4. Purification of Bax from murine thymocytes. *Methods:* Bax was purified from murine thymocyte soluble protein fraction by heparin affinity, TMAE anion exchange, and immunoaffinity chromatography (with anti-murine Bax 5B7 monoclonal antibody). The protein samples were analyzed by SDS-polyacrylamide gel electrophoresis (top) and Western blotting (bottom) analysis with anti-Bax monoclonal antibody 2C8. *Lane a*, soluble murine thymocyte extract; *lane b*, flow-through from heparin column; *lane c*, flow-through from TMAE anion-exchange column; *lane d*, eluant from TMAE column; *lane e*, antibody column eluant.

that Bax, in the presence of nonionic detergents, can undergo a conformational change *(19)*. This can lead to the exposure of its 6A7 antibody epitope and Bax homodimerization and heterodimerization with Bcl-2 and Bcl-X_L (Table 2). This nonionic detergent-induced Bax dimerization could significantly decrease Bax interaction with its actual binding partners if inappropriate detergents are used during membrane solubilization. Due to these factors, identification of Bax-binding partners by immunoprecipitation should be carried

Table 2
Bax Dimer Formation and the Exposure of the
6A7 Antibody Epitope in the Presence of Different Detergents

Detergent	Heterodimerization		Homodimerization	6A7 epitope exposure
	Bcl-2	Bcl-XL		
None[a]	N/A	−	−	−
Triton X-100	++	+++	+++	+++
Nonidet P-40	++	+++	+++	+++
Triton X-114	+++	+++	+++	++
Chaps	−	−	−	−
Octylglucoside	+	+++	−	−
Dodecyl maltoside	−	++	+++	+

Methods: Immunoprecipitation of Bax was carried out to determine Bax dimerization state and exposure of the 6A7 antibody epitope from either soluble protein extract or detergent-solubilized murine thymocyte lysate.
[a]Studies done with soluble protein extract.

out with due diligence, and subsequent interpretation must be examined with great care. Finally, one would need to supplement the immunoprecipitation results with functional studies to prove that this protein interaction modulates Bax function in apoptosis.

Taken together, the above-described techniques could prove useful in studying the role of Bax in mediating cell death in different cell types undergoing apoptosis.

References

1. Oltvai, Z. N., Milliman, C. L., and Korsmeyer, S. J. (1993) Bcl-2 heterodimerizes in vivo with a conserved homolog, Bax, that accelerates programmed cell death. *Cell* **74,** 609–619.
2. Yang, E. and Korsmeyer, S. J. (1996) Molecular thanatopsis: a discourse on the Bcl2 family and cell death. *Blood* **88,** 386–401.
3. Knudson, C. M., Tung, K. S. K., Tourtellotte, W. G., Brown, G. A. J., and Korsmeyer, S. J. (1995) Bax-deficient mice with lymphoic hyperplasia and male germ cell death. *Science* **270,** 96–99.
4. Rodriguez, I., Ody, C., Araki, K., Garcia, I., and Vassalli, P. (1997) An early and massive wave of germinal cell apoptosis is required for the development of functional spermatogenesis. *EMBO J.* **16,** 2262–2270.
5. Krajewski, S., Mai, J. K., Krajewska, M., Sikorska, M., Mossakowski, M. J., and Reed, J. C. (1995) Upregulation of Bax protein levels in neurons following cerebral ischemia. *J. Neurosci.* **15,** 6364–6376.
6. Cheng, W., Kajstura, J., Nitahara, J. A., et al. (1996) Programmed myocyte cell death affects the viable myocardium after infarction in rats. *Exp. Cell Res.* **226,** 316–327.

7. Misao, J., Hayakawa, Y., Ohno, M., Kato, S., Fujiwara, T., and Fujiwara, H. (1996) Expression of bcl-2 protein, an inhibitor of apoptosis, and Bax, an accelerator of apoptosis, in ventricular myocytes of human hearts with myocardial infarction. *Circulation* **94,** 1506–1512.

8. Rampino, N., Yamamoto, H., Ionov, Y., et al. (1997) Somatic frameshift mutations in the BAX gene in colon cancers of the microsatellite mutator phenotype. *Science* **275,** 967–969.

9. Yin, X.-M., Oltvai, Z. N., and Korsmeyer, S. J. (1994) BH1 and BH2 domains of Bcl-2 are required for inhibition of apoptosis and heterodimerization with Bax. *Nature* **369,** 321–323.

10. Zha, H., C. Aime-Sempe, Sato, T., and Reed, J. C. (1996) Proapoptotic protein Bax heterodimerizes with Bcl-2 and homodimerizes with Bax via a novel domain (BH3) distinct from BH1 and BH2. *J. Biol. Chem.* **271,** 7440–7444.

11. Hockenbery, D., Nunez, G., Milliman, C., Schreiber, R. D., and Korsmeyer, S. J. (1990) Bcl-2 is an inner mitochondrial membrane protein that blocks programmed cell death. *Nature* **348,** 334–336.

12. Krajewski, S., Tanaka, S., Takayama, S., Schibler, M. J., Fenton, W., and Reed, J. C. (1993) Investigation of the subcellular distribution of the bcl-2 oncoprotein: residence in the nuclear envelope, endoplasmic reticulum, and outer mitochondrial membranes. *Cancer Res.* **53,** 4701–4714.

13. Suzuki, M., Youle, R. J., and Tjandra, N. (2000) Structure of Bax. Coregulation of dimer formation and intracellular localization. *Cell* **103,** 645–654.

14. Muchmore, S. W., Sattler, M., Liang, H., et al. (1996) X-ray and NMR structure of human Bcl-XL, an inhibitor of programmed cell death. *Nature* **381,** 335–341.

15. Kagan, B. L., Finkelstein, A., and Colombini, M. (1981) Diphtheria toxin fragment forms large pores in phospholipid bilayer membranes. *Proc. Natl. Acad. Sci. USA* **78,** 4950–4954.

16. Hu, V. W. and Holmes, R. K. (1984) Evidence for direct insertion of fragments A and B of diphtheria toxin into model membranes. *J. Biol. Chem.* **259,** 12,226–12,233.

17. Zhan, H., Oh, K. J., Shin, Y.-K., Hubbell, W. L., and Collier, R. J. (1995) Interaction of the isolated transmembrane domain of diphtheria toxin with membranes. *Biochemistry* **34,** 4856–4863.

18. Hsu, Y.-T., Wolter, K. G., and Youle, R. J. (1997) Cytosol-to-membrane redistribution of Bax and Bcl-XL during apoptosis. *Proc. Natl. Acad. Sci. USA* **94,** 3668–3672.

19. Hsu, Y.-T. and Youle, R. Y. (1998) Bax in murine thymus is a soluble monomeric protein that displays differential detergent-induced conformations. *J. Biol. Chem.* **273,** 10,777–10,783.

20. Gross, A., Jockel, J., Wei, M. C., and Korsmeyer, S. J. (1998) Enforced dimerization of Bax results in its translocation, mitochondrial dysfunction and apoptosis. *EMBO J.* **17,** 3878–3885.

21. Wolter, K. G., Hsu, Y.-T., Smith, C. L., Nechushtan, A., Xi, X.-G., and Youle, R. J. (1997) Movement of Bax from the cytosol to mitochondria during apoptosis. *J. Cell Biol.* **139,** 1281–1292.

22. Saikumar, P., Dong, Z., Patel, Y., et al. (1998) Role of hypoxia-induced Bax translocation and cytochrome c release in reoxygenation injury. *Oncogene* **17,** 3401–3415.

23. Zhang, H., Heim, J., and Meyhack, B. (1998) Redistribution of Bax from cytosol to membranes is induced by apoptotic stimuli and is an early step in the apoptotic pathway. *Biochem. Biophys. Res. Commun.* **251,** 454–459.

24. Finucane, D. M., Bossy-Wetzel, E., Waterhouse, N. J., Cotter, T. G., and Green, D. R. (1999) Bax-induced caspase activation and apoptosis via cytochrome c release from mitochondria is inhibitable by Bcl-XL. *J. Biol. Chem.* **274,** 2225–2233.

25. Khaled, A. R., Kim, K., Hofmeister, R., Muegge, K., and Durum, S. K. (1999) Withdrawal of IL-7 induces Bax translocation from cytosol to mitochondria through a rise in intracellular pH. *Proc. Natl. Acad. Sci. USA* **96,** 14,476–14,481.

26. Murphy, K. M., Streips, U. N., and Lock, R. B. (1999) Bax membrane insertion during Fas(CD95)-induced apoptosis precedes cytochrome c release and is inhibited by Bcl-2. *Oncogene* **18,** 5991–5999.

27. Putcha, G. V., Deshmukh, M., and Jr., E. M. J. (1999) BAX translocation is a critical event in neuronal apoptosis: regulation by neuroprotectants, BCL-2, and caspases. *J. Neurosci.* **19,** 7476–7485.

28. Ghatan, S., Larner, S., Kinoshita, Y., et al. (2000) p38 MAP kinase mediates bax translocation in nitric oxide-induced apoptosis in neurons. *J. Cell Biol.* **150,** 335–348.

29. Nechushtan, A., Smith, C. L., Hsu, Y.-T., and Youle, R. J. (1999) Conformation of the Bax C-terminus regulates subcellular location and cell death. *EMBO J.* **18,** 2330–2341.

30. Manon, S., Chaudhuri, B., and Guerin, M. (1997) Release of cytochrome c and decrease of cytochrome c oxidase in Bax-expressing yeast cells, and prevention of these effects by coexpression of Bcl-XL. *FEBS Lett.* **415,** 29–32.

31. Jurgensmeier, J. M., Xie, Z., Deveraux, Q., Ellerby, L., Bredesen, D., and Reed, J. C. (1998) Bax directly induces release of cytochrome c from isolated mitochondria. *Proc. Natl. Acad. Sci. USA* **95,** 4997–5002.

32. Pastorino, J. G., Chen, S.-T., Tafani, M., Snyder, J. W., and Farber, J. L. (1998) The overexpression of Bax produces cell death upon induction of the mitochondrial permeability transition. *J. Biol. Chem.* **273,** 7770–7775.

33. Schlesinger, P. H., Gross, A., Yin, X. M., et al. (1997) Comparison of the ion channel characteristics of proapoptotic Bax and antiapoptotic Bcl-2. *Proc. Natl. Acad. Sci. USA* **94,** 11,357–11,362.

34. Basanez, G., Nechushtan, A., Drozhinin, O., et al. (1999) Bax, but not Bcl-XL, decreases the lifetime of planar phospholipid bilayer membranes at subnanomolar concentrations. *Proc. Natl. Acad. Sci. USA* **96,** 5492–5497.

35. Shimizu, S., Narita, M., and Tsujimoto, Y. (1999) Bcl-2 family proteins regulate the release of apoptogenic cytochrome c by the mitochondrial channel VDAC. *Nature* **399,** 483–487.

36. Marzo, I., Brenner, C., Zamzami, N., et al. (1998) Bax and adenine nucleotide translocator cooperate in the mitochondrial control of apoptosis. *Science* **281,** 2027–2031.

37. Li, P., Nijhawan, D., Budihardjo, I., et al. (1997) Cytochrome c and dATP-dependent formation of Apaf-1/caspase-9 complex initiates an apoptotic protease cascade. *Cell* **91,** 479–489.

38. Murphy, K. M., Ranganathan, V., Farnsworth, M. L., Kavallaris, M., and Lock, R. B. (2000) Bcl-2 inhibits bax translocation from cytosol to mitochondria during drug-induced apoptosis of human tumor cells. *Cell Death Differ.* **7,** 102–111.

Assessing Cytochrome-*c* Release from Mitochondria

Douglas W. Ethell and Douglas R. Green

1. Introduction

In recent years substantial evidence has accumulated demonstrating a central regulatory role for mitochondria in cell death *(1)*. More recently, mechanisms of cell death, through apoptosis, have been elucidated and increasingly point toward mitochondria as the gatekeepers of apoptosis. The morphological changes that define apoptosis can be attributed to the actions of a family of cysteinyl-dependent aspartate specific proteinases, called caspases *(2)*. Members of this highly conserved protein family are expressed by all metazoan cells as zymogen proforms, which become activated by specific cleavage events that are usually mediated by other caspases. That is, activated caspases can cleave and activate other caspases, creating a cascade. The first caspases to become activated in this enzymatic cascade are called apical, or initiator caspases, and contain large prodomains that facilitate their recruitment to specialized macromolecular complexes. Once recruited to these complexes, apical caspases become activated and this can quickly lead to death of the cell through apoptosis. Therefore, the formation and regulation of these complexes are carefully regulated. Of the two apical caspase cascades identified thus far, mitochondria play a central role in one pathway and an amplification role in the other.

One source of apical caspase activation is a cytoplasmic macromolecular complex called the apoptosome. Formation and activation of the apoptosome requires procaspase-9, Apaf-1, dATP/ATP from the cytoplasm, and cytochrome-*c* from the intermembrane space of mitochondria *(3,4)*. Release of cytochrome-*c* and other

From: *Neuromethods, Vol. 37: Apoptosis Techniques and Protocols, 2nd Ed.*
Edited by: A. C. LeBlanc @ Humana Press, Inc., Totowa, NJ

proapoptotic components from mitochondria is a carefully regulated step in apoptosis *(5)*. The involvement of cytochrome-*c* in caspase activation sparked the initial interest in this area, although it has since been shown that release of SMAC/DIABLO from mitochondria can affect caspase activity by inhibition of apoptosis proteins (IAPs), causing a disinhibition of caspases *(6,7)*. More recently, it has been suggested that apoptosis-inducing factor (AIF) is released from mitochondria and may regulate a new cell death pathway *(8)*.

Cytochrome-*c* plays a key role in oxidative phosphorylation by shuttling electrons between complexes III and IV of the electron-transport chain. Although mitochondrial in location, it is encoded on the cellular genome and translated in the cytoplasm as apocytochrome-*c*. This immature form of cytochrome-*c* is not capable of playing a role in the electron transport chain or initiating apoptosome formation. Apocytochrome-*c* is imported to the intermembrane space through the outer mitochondrial membrane (OMM). Once inside the OMM, apocytochrome-*c* is covalently attached to a heme group to form mature holocytochrome-*c*. Holocytochrome-*c* is more globular and is trapped between the mitochondrial membranes. However, proapoptotic stimuli can cause the release of cytochrome-*c* and other proteins of the intermembrane space into the cytoplasm by an as yet unknown mechanism. Once released from mitochondria, cytochrome-*c* can interact with Apaf-1, which makes Apaf-1 capable of recruiting procaspase-9, forming an apoptosome. Association with Apaf-1 is required for caspase-9 activity, which can then cleave caspase-3 *(9)*. Once activated, caspase-3 can cleave more caspases and other substrates throughout the cell.

Studies involving the regulation and activation of the mitochondria–cytochrome-*c* pathway need to assess several key steps to establish activation. These steps can include, but are not limited to,

1. Subcellular localization of cytochrome-*c*—mitochondrial vs cytoplasmic
2. Caspase activation—caspase-3 cleavage and/or activity
3. Determination of apoptosis—morphological and biochemical

The subcellular localization of cytochrome-*c* can be accomplished using immunocytochemistry or biochemistry. Each approach offers many advantages and can be used to address the three points above. Ideally, both approaches can be employed; however, due to the technical difficulties, it may be preferable to concentrate on doing one well.

2. Immunostaining
for Cytochrome-c and Bcl-2 Family Proteins

Many neuronal cultures are mixed populations of neurons and glia, so cytochrome-*c* release in neurons may be more easily assessed with immunostaining. Further, the same techniques can be used to study whether there are changes in the subcellular localization of proapoptotic Bcl-2 family members or other proteins of interest in cell death research, such as SMAC and AIF. In addition, there are no limitations on the selection of neurons for use in immunostaining studies, as the cells remain on their culture substrate.

2.1. Advantages

The variety of antibodies specific for apoptotic machinery allows for a thorough analysis of apoptosis, including nuclear morphology and caspase-3 cleavage. 4'6-diamine-2 phenylindol (DAPI)-stained nuclei fluoresce (blue) with UV excitation, allowing triple staining with FITC (green) and Texas Red (red) conjugated secondary antibodies. Further, phase contrast can be used to assess morphological changes. Finally, terminal deoxynucleotidyl transferase-mediated UTP nick and labeling (TUNEL) can be used to assess DNA fragmentation, indicative of apoptosis.

2.2. Disadvantages

1. Requires access to a confocal microscope or a good fluorescence microscope.
2. Fluorescence will fade.

2.3. Reagents and Solutions

See Table 1.

2.4. Antibodies for Immunostaining

See Table 2.

2.5. Paraformaldehyde Fixative

1. Mix 4.0 g paraformaldehyde in ~60 mL H_2O with one drop of 10 N NaOH.
2. Warm to no more than 60°C, on a heating stir plate.

Table 1
Reagents and Solutions for Immunostaining

Reagent	Catalog number	Vendor
Bovine serum albumin (BSA)	BP1605-100	Fisher
Chambered cover glass	155409	Nalge-NUNC
Cover slips	12-545-82	Fisher
CytosealXYL	8312-4	Stephens Scientific (Kalamazoo, MI)
DAPI	D8417	Sigma
Dimethyl sulfoxide (DMSO)	D2650	Sigma
Normal goat serum (NGS)	G9023	Sigma (St. Louis, MO)
Paraformaldehyde	T353-500	Fisher (Fair Lawn, NJ)
10X phosphate-buffered saline (PBS)	70011-044	Gibco-BRL (Gaithersburg, MD)
Tween-20	BP337-500	Fisher
Vectashield mounting medium	H-1000	Vector labs (Burlingame, MA)
Vectashield + DAPI	H-1200	Vector labs

Table 2
Antibodies for Immunostaining

Antigen	Dilution	Catalog number	Vendor
Cytochrome-*c*	1/1000	556432	BD-Pharmingen (San Diego, CA)
Bax	1/1000	6A7	BD-Pharmingen
Cleaved caspase-3	1/100	9661S	Cell Signaling Tech. (Beverly, MA)
SMAC/DIABLO	1/100	804333C100	Alexis (San Diego, CA)
FITC-conj., anti-rabbit IgG		F9887	Sigma
Texas Red-conj., anti-mouse IgG		715-076150	Jackson Immuno

3. Add 10 mL 10X PBS.
4. Cool on ice for 5 min.
5. Adjust pH to 7.4 with HCl.
6. Add water to a final volume of 100 mL.
7. Run through a bottle-top filter with vacuum.
8. Fixative should be made fresh and stored on ice, or aliquots kept at $-20°C$.

Caution: Paraformaldehyde is very toxic and carcinogenic. Weigh powder and work with solutions in a fume hood.

2.6. PBS-Tween (PBST)

PBST: 1X PBS + 1/1000 vol of Tween-20.

2.7. Blocking Buffer

1. Completely dissolve 3.0 g of bovine serum albumin into 60 mL of dH$_2$O with stirring. Add a little more every few minutes until it is completely dissolved.
2. Add 3 mL of normal goat serum (NGS).
3. Add 10 mL of 10X PBS (pH 7.4).
4. Bring to 100 mL final volume.
5. Aliquot and store at –20°C.

2.8. Primary and Secondary Antibody Incubation Buffer

1. 4 mL blocking buffer (*see* Subheading 2.7.).
2. 8 mL 1X PBS.
3. 12 μL Tween-20.
4. Make fresh every time.

2.9. Fixation and Immunostaining of Neuronal Cultures

Once the cultures are ready for analysis:

1. Gently remove the medium (always use a pipetman, not aspiration).
2. Gently rinse 3X with PBS.
3. Overlay with fresh paraformaldehyde fixative (*see* Subheading 2.5.).
4. Leave 15–20 min at room temperature.
5. Rinse 3X with PBS.
6. Permeabilize the cells for 5 min with 0.2% Triton X-100/0.25% sarcosine (w/v) in PBS.
7. Rinse 3X with PBS.
8. Block nonspecific binding with 3% NGS/3% BSA/PBST for 30–60 min at room temperature.
9. Dilute primary antibodies in 1% NGS/1% BSA in PBS + 0.1% Tween-20.
10. Incubate primary for 1–2 h at room temperature, or overnight at 4°C.

11. Remove primary antibody and rinse 3X for 10 min each time with PBST.
12. Dilute fluorescently labeled secondary antibody in 1% NGS/ 1% BSA/PBST, 1 h at room temperature.
13. Rinse 3X 10 min each time with PBST.
14. Rinse with PBS (no Tween-20).

We recommend using chambered cover glasses, which make immunostaining very easy because of the wells. At this point in the protocol, leave the chambers on the cover glasses and simply cover the cells with PBS or mounting medium; trying to remove the wells from a chambered cover glass is unnecessary, and they break easily. If the high cost of these chambers is prohibitive, then treated cover slips cultured in 24-well plates will work just as well, but will require mounting before microscopy. Alternatively, chambered slides can be used and mounted onto large cover slips when done. Thin cover glass is required to take full advantage of the fluorescence.

2.10. Mounting Cover Slips

Prepare everything first!

1. Remove cover slips from PBS using blunt tweezers. Hold vertically and touch the edge to a Kimwipe to remove most of the solution. Lay on a Kimwipe, cell side up, for 1–2 min. The cover slips should be dry, but not bone-dry.
2. Place a drop of mounting medium onto a clean glass slide.
3. Invert the cover slip and touch the edge of the cover slip to the drop.
4. Gently lower the cover slip on to the drop so that no air bubbles are trapped. If you get air, then try to remove them by gently pushing on the cover slip with blunt tweezers. It is often best to use a little extra mounting medium and remove the excess at the edge of the cover glass with a 10-µL pipet.
5. Seal the cover glass in place with CytosealXYL.
6. Mounting chambered slides is similar: remove the chambers, and mount 0.17-mm cover glass onto the slide.
7. Store fluorescently labeled samples in the dark at 4°C.

2.11. Mitochondrial Co-localization

Co-localization of cytochrome-*c*, Bcl-2, Bax, or other Bcl-2 family members with mitochondria can be done in fixed cells using Mitotracker Green (Molecular Probes, Eugene, OR), which does not depend on inner mitochondrial membrane (IMM) potential ($\Delta\Psi_m$) for mitochondrial localization. Incubation of fixed cultures with 150 nM Mitotracker Green for 30 min at room temperature is sufficient to label mitochondria. Alternatively, Mitotracker Red (150 nM) can be added to culture media 10–15 min prior to fixation. However, Mitotracker Red will only label mitochondria with $\Delta\Psi_m$ across the IMM, and it is toxic to cells when used for longer times. That is, Mitotracker Red will not label the mitochondria in late apoptotic cells or those that have died by necrosis. Alternatively, one can use double immunostaining for antigens that are retained by mitochondria in apoptotic cells, such as COX IV.

2.12. Staining for Nuclear Morphology

The nuclear morphology of fixed, permeabilized, and immunostained cultures can be visualized by staining DNA with DAPI using either of two methods. First, mounting medium that contains DAPI can be used. Second, DAPI that has been solubilized to 2.5 mg/mL in DMSO can be diluted 1/1000 in PBS and overlaid on the cells for 20–30 min at room temperature. Cells should be rinsed 3X with PBS afterward.

3. Biochemical Localization of Cytochrome-c

In biochemical localization of cytochrome-*c* in cytosolic and mitochondrial fractions, cells are isolated and disrupted, without damaging mitochondria (modified from [10]). Centrifugation is used to separate cytosolic and mitochondria fractions. Western blotting of those fractions reveals relative concentrations of cytochrome-*c* in each.

In selecting neurons for use in cytochrome-*c* studies, one must consider the subsequent biochemical steps for the purification of intact mitochondria. Sympathetic and cortical neurons can be grown on polyornithine and laminin. These neurons can be easily dislodged from the dish and pelleted. Also, cortical and sympathetic neurons have good-sized cytoplasms, which is desirable for homogenization. Optimally, the plasma membranes should be disrupted while keeping as many nuclei and mitochondria intact as possible. Conversely, cerebellar neuron cultures are usually grown on poly-D-lysine or poly-L-lysine, both of which grip the cells so tightly as to make their removal difficult. Scraping neurons from a poly-D-lysine coated dish shears the cells and makes for difficult biochemistry. Further, granular neurons have a large nucleus and relatively small cytoplasm, which also makes for difficult homogenization.

3.1. Advantages

When done correctly, the findings are definitive and highly reproducible. Also, the mitochondrial translocation of proapoptotic Bcl-2 family proteins, such as Bax, can be assessed with the same fractions. These fractions can also be probed for many proteins that are cleaved in apoptosis. In addition, the cytoplasmic extracts can be assayed directly for caspase activity using fluorescent substrates.

3.2. Disadvantages

Consistency and attention to detail make this process technically demanding.

3.3. Reagents and Solutions

Reagent	Catalog number	Vendor
BCA protein assay	23225	Pierce (Rockford, IL)
Complete protease inhibitor cocktail	1697498	Roche (Indianapolis, IN)
2-mL Dounce homogenizer with tight-fitting glass pestle (B)		Wheaton
HybondECL	RPN 2020D	Amersham (Piscataway, NJ)
Precast minigels (4–15% gradient)	161-1158	Biorad (Hercules, CA)
Supersignal chemiluminescent reagent	34075	Pierce

3.4. Antibodies for Western Blotting

Antigen	Dilution	Catalog number	Vendor
Cytochrome-*c*	1/1000	556433	BD-Pharmingen
COX IV	1/1000	A-3431	Molecular Probes (Eugene, OR)

3.5. Homogenization Buffer

1. Dissolve 8.6 g of sucrose in 50 mL dH$_2$O.
2. Dissolve 2 tablets of complete protease inhibitor cocktail.
3. Add 2 mL of 1 M HEPES (pH 7.4).
4. Add 1 mL of 1 M KCl.

5. Add 400 µL of 0.5 M EDTA.
6. Add 100 µL of 1 M MgCl$_2$.
7. Add 100 µL of 1 M dithiothreitol (DTT).
8. Bring to 100 mL final volume with dH$_2$O.
9. Filter-sterilize and keep at 4°C for up to 1 mo.

3.6. Mitochondria Buffer

1. Dissolve 2 tablets of complete protease inhibitor cocktail in 50 mL dH$_2$O.
2. Add 5 mL of 1 M HEPES (pH 7.4).
3. Slowly add 1 mL of NP-40.
4. Add 10 mL glycerol.
5. Add 100 µL of 1 M EDTA.
6. Add 20 µL of 1 M DTT.
7. Bring to 100 mL final volume with dH$_2$O.
8. Filter-sterilize and keep at 4°C for up to 1 mo.

3.7. 5X Protein Sample Buffer

1. Place 50 mL dH$_2$O in a 100-mL beaker, with stir bar.
2. Add 25 mL of 1 M Tris-HCl (pH 6.8).
3. Add 10 g SDS.
4. Slowly add 10 mL glycerol.
5. Add 0.1 g bromophenol blue.
6. Bring to 100 mL final volume with dH$_2$O.
7. Store at room temperature with stir bar.
8. Prior to use, mix again with stirring.
9. Place 950 µL in an Eppendorf tube.
10. Add 50 µL of 1 M DTT.
11. Close lid and vortex to mix.

3.8. Transfer Buffer

1. Dissolve 5 g of Tris-HCl into 1400 mL dH$_2$O.
2. Add 23 g glycine.
3. Bring to final volume with 1600 mL dH$_2$O.
4. Add 400 mL methanol.
5. Cool before using, keep at 4°C.

3.9. Blotto (Blocking Buffer)

1. Dissolve 3 g of bovine serum albumin (BSA) into 60 mL of dH$_2$O, add slowly with high-speed mixing.

2. Add 3 g dried nonfat milk.
3. Add 10 mL 10X PBS.
4. Add 100 µL Tween-20.
5. Add dH$_2$O to 100-mL final volume.

3.10. Antibody (Ab) Buffer

1. Dilute 1 part blotto with 2 parts PBST.

3.11. Stripping Buffer

1. Place 50 mL dH$_2$O into a 100-mL flask, with stir bar.
2. Add 7 mL of 1 M Tris-HCl (pH 6.8).
3. Add 2 g SDS.
4. Add 1 mL of 2-mercaptoethanol.
5. Bring to 100-mL final volume with dH$_2$O.
6. Store at room temperature.

3.12. Preparation of Cytosolic and Mitochondria Fractions

These procedures should be done with ice-cold buffers and kept on ice throughout.

1. Start with 10^7 neurons/sample and scale down as appropriate. Cultures should be rinsed gently 3X with cold PBS. Neurons cultured on polylysine + laminin can be easily dislodged with a pipetman, which is best as it helps maintain cell integrity. Alternatively, if the cells do not come off easily, they can be gently dislodged by scraping with a rubber policeman. In either case the dish should be subsequently washed with cold PBS and that pooled to get all of the cells.
2. Pellet neurons at 500g for 5 min.
3. Rinse with 10 mL cold 1X PBS; carefully remove all PBS.
4. Resuspend pellet in 500 µL homogenization buffer, leave on ice for 20–25 min, flick tube every 2 min.
5. Prerinse a cooled 2.0-mL Dounce homogenizer with ice-cold homogenization buffer to prevent adsorption of protein to the sides. Do not rinse with just dH$_2$O, as the residue will change the osmolarity of the homogenization buffer and give spurious results.
6. Homogenize sample with 30–50 strokes, continuous motion, keeping the pestle inside the buffer to reduce bubbling. The number of strokes should be worked out in pilot studies with

that particular homogenizer and the specific cells used. Too much homogenization will rupture nuclei and mitochondria, whereas too little will not break up all of the cells.

7. Transfer homogenate to a precooled Eppendorf tube, close, and keep on ice.
8. Rinse the homogenizer with 2 mL of cold homogenization buffer.
9. Proceed with next sample, be as quick as possible, but don't let a homogenized sample sit on ice for more than 15–20 min.
10. When all samples are done, spin down large debris at 800*g* for 10 min at 4°C.
11. Carefully remove and keep supernatant, leaving some at the bottom so as not to contaminate with debris from the pellet.
12. Centrifuge supernatant in a 4°C Eppendorf centrifuge at top speed (~22,000*g*) for 15 min.
13. Carefully remove the supernatant (this is the "cytosolic fraction"), making sure not to contaminate it with any of the pellet. Leave the last few microliters of supernatant. Store the cytosolic fractions at −80°C until ready.
14. Remove the remaining few microliters of supernatant and discard; keep the pellet.
15. Resuspend the pellet in 100 µL of cold mitochondria buffer and incubate on ice for 20 min, flick tube every 2 min.
16. Centrifuge in a 4°C Eppendorf centrifuge at top speed (~22,000*g*) for 15 min.
17. Carefully remove the supernatant, being careful not to contaminate with the pellet.
18. This supernatant is the "mitochondria fraction" and should be stored at −80°C.

3.13. Electrophoresis and Blotting

1. Quantitate protein concentrations with BCA protein assay, using homogenization and mitochondria buffers to baseline respective samples.
2. Mix 10–50 µg of protein with 10 mL of 5X sample buffer.
3. Boil 5–10 min.
4. Cool on ice, quick spin to get all solution to the bottom.
5. Load samples onto 4–15% gradient gels, use prestained markers.
6. Electrophorese until bromophenol blue dye nears the bottom of the gel.

7. Remove gel from glass plates and quickly rinse with transfer buffer.
8. Electroblot gel onto HybondECL membrane (160 mA, overnight).
9. Remove membrane from cassette, confirm transfer of prestained markers.

3.14. Western Blotting

1. Rinse membrane 3X for 10 min each time in 100 mL PBST.
2. Block nonspecific binding in 100 mL blotto for 2 h at room temperature.
3. Mix primary antibody in Ab buffer; for cytochrome-c add 6 µL to 6 mL.
4. Seal membrane in a freezer bag with the primary Ab solution, remove all air bubbles.
5. Incubate on an orbital rotator at 4°C overnight.
6. Remove membrane from bag.
7. Rinse membrane 3X for 10 min each time in 100 mL PBST.
8. Dilute secondary antibody with Ab buffer.
9. Seal membrane and secondary Ab solution in a new freezer bag.
10. Place on an orbital rotator for 2 h at room temperature.
11. Remove membrane from bag.
12. Rinse membrane 3X for 10 min each time in 100 mL PBST.
13. Rinse for 5 min in 100 mL PBS (no Tween-20).
14. While in final rinse, prepare chemiluminescent reagents (West Pico).
15. Place membrane on an overhead transparency.
16. Cover the entire membrane with chemiluminescent reagent, let sit for 60 s.
17. Place the membrane on a piece of transparent sheet protector.
18. Use blot to expose film as required.
19. If the signal is not strong enough even after a 15–20 min exposure, rinse membrane with PBS and try again with the stronger chemiluminescent reagent (West Dura).

3.15. Stripping

The same membrane should be reprobed for actin (cytosolic fraction) or cytochrome oxidase subunit IV (COX IV) (mitochondria fraction) to confirm consistent loading of samples. Seal the membrane in a freezer bag with 10 mL of stripping buffer and immerse in a 55°C water bath for 5 min. Remove membrane and rinse for 2 h in dH$_2$O, with constant changes of water.

4. Questions That Can Be Answered With These Techniques

1. *Has cytochrome-c been released from mitochondria?* Immunostaining for cytochrome-*c* should be punctate if it is in the mitochondria, but diffuse if it has been released. Blots of mitochondria fractions from control and treated samples will be the same if there has not been release, but different if there has been release. Conversely, cytosolic fractions of viable cells will not contain cytochrome-*c*, whereas cytosolic fractions from apoptotic cells will. It is important to note that "cytochrome-*c* release" is most likely indicative of a general permeabilization of the mitochondrial outer membrane.

2. *Has Bax translocated to mitochondria?* Immunostaining will show diffuse staining for Bax prior to mitochondrial insertion, and punctate staining thereafter. Using the biochemical approach, Bax should be found in the cytosolic fraction of nonapoptotic cells, and the mitochondria fraction of apoptotic cells.

3. *Have caspases been activated?* Immunostaining with antibody to active caspase-3 will reveal which cells have undergone apoptosis. Blotting cytosolic fractions with caspase-3-specific antibodies will show how much full-length vs cleaved caspase-3 is present. Also, the cytosolic fractions can be used for DEVDase assays to measure caspase activity.

4. *Have substrates characteristic of apoptosis been cleaved?* Nuclear morphology can be seen with DAPI staining. Blotting for classic caspase substrates such as PARP will establish if cleaved caspase-3 is fully active, and not been inhibited by IAPs of other antiapoptotic mechanisms.

References

1. Green, D. R. (2000) Apoptotic pathways: paper wraps stone blunts scissors. *Cell* **102(1),** 1–4.
2. Thornberry, N. A. and Lazebnik, Y. (1998) Caspases: enemies within. *Science* **281(5381),** 1312–1316.
3. Zou, H., Henzel, W. J., Liu, X., Lutschg, A., and Wang, X. (1997) Apaf-1, a human protein homologous to *C. elegans* CED-4, participates in cytochrome c-dependent activation of caspase-3. *Cell* **90(3),** 405–413.

4. Bossy-Wetzel, E., Newmeyer, D. D., and Green, D. R. (1998) Mitochondrial cytochrome c release in apoptosis occurs upstream of DEVD-specific caspase activation and independently of mitochondrial transmembrane depolarization. *EMBO J.* **17(1)**, 37–49.

5. Gross, A., McDonnell, J. M., and Korsmeyer, S. J. (1999) BCL-2 family members and the mitochondria in apoptosis. *Genes Dev.* **13(15)**, 1899–1911.

6. Du, C., Fang, M., Li, Y., Li, L., and Wang, X. (2000) Smac, a mitochondrial protein that promotes cytochrome c-dependent caspase activation by eliminating IAP inhibition. *Cell* **102(1)**, 33–42.

7. Verhagen, A. M., Ekert, P. G., Pakusch, M., et al. (2000) Identification of DIABLO, a mammalian protein that promotes apoptosis by binding to and antagonizing IAP proteins. *Cell* **102(1)**, 43–53.

8. Joza, N., Susin, S. A., Daugas, E., et al. (2001) Essential role of the mitochondrial apoptosis-inducing factor in programmed cell death. *Nature* **410(6828)**, 549–554.

9. Stennicke, H. R. and Salvesen, G. S. (1999) Catalytic properties of the caspases. *Cell Death Differ.* **6(11)**, 1054–1059.

10. Bossy-Wetzel, E. and Green, D. R. (2000) Assays for cytochrome c release from mitochondria during apoptosis. *Meth. Enzymol.* **322**, 235–242.

Monitoring Caspases in Neuronal Cell Death

Younes Bounhar, Omar Tounekti, and Andréa C. LeBlanc

1. Introduction

Apoptosis is a physiological process that contributes to the establishment and homeostasis of the nervous system. For example, neurons that fail to make the proper connections with their postsynaptic targets die naturally during development due to the lack of sufficient trophic support *(1)*. Cell demise is operated in a well-ordered fashion and culminates in the activation of a network of specific proteases, the caspases that execute the death program. Once activated by the apoptotic signal, caspases cleave a myriad of cellular target proteins in a highly specific fashion. As a consequence, the death signal is amplified, and the committed cell systematically dismantled *(2)*. We describe here a practical guide to investigate the involvement of caspases in neuronal cell death.

2. Caspases

Caspases are a unique family of cysteinyl aspartate proteases that display an absolute requirement for aspartate for proteolytic activity *(3)*. To date, 14 mammalian caspases have been identified, 11 of which are found in humans (Table 1). Recently, caspase homologs lacking caspase protease activity have been identified in humans, *Caenorhabditis elegans* and *Dictyostelium discoideum* (the paracaspases), and in *Arabidopsis thaliana* (the metacaspases) *(4)*. Caspases share homology at the amino acid and at the structural level. Caspases are synthesized as inactive zymogens. They possess

From: *Neuromethods, Vol. 37: Apoptosis Techniques and Protocols, 2nd Ed.*
Edited by: A. C. LeBlanc @ Humana Press, Inc., Totowa, NJ

Table 1
General Characteristics of Mammalian Caspases

Caspase	Group	Function	Zymogen (kDa)	Large subunit	Small subunit	Preferred substrate
Caspase-1	I	Cytokine	45	20	10	(W/Y/F)EHD
Caspase-2	II	Initiator	51	20	12	VDVAD[a]/ DXXD[b]
Caspase-3	II	Effector	32	17	12	DEXD
Caspase-4	I	Cytokine	43	20	10	(W/L/F)EHD
Caspase-5	I	Cytokine	48	20	10	(W/L/F)EHD
Caspase-6	III	Effector	34	18	11	(V/T/I)EXD
Caspase-7	II	Effector	35	20	12	DEXD
Caspase-8	III	Initiator	55	18	11	(L/V/D)EXD
Caspase-9	III	Initiator	45	17	10	(I/V/L)EHD
Caspase-10	III	Initiator	55	17	12	DEVD
mCaspase-11[c]	ND	Cytokine	42	20	10	ND
mCaspase-12[c]	ND	Cytokine	50	20	10	ND
Caspase-13	ND	Cytokine	43	20	10	ND
mCaspase-14[c]	ND	Cytokine	30	20	10	ND

[a]From ref. *10*.
[b]From ref. *3*.
[c]mCaspase: murine caspase.
Adapted from ref. *11*.

an N-terminal prodomain, which varies from 20 to over 200 residues in length, a large subunit (~20 kDa) comprising the QACXG consensus catalytic site, and a C-terminal small subunit (~10 kDa) (Table 1). The active caspase is a tetramer formed by two heterodimers composed of a large and a small subunit *(5–8)*.

Despite their common requirement for Asp at P1 and a high degree of homology in the substrate-binding sites, caspases display two major differences. First, several large studies conducted on 10 caspases *(3,9,10)* have shown that caspases have discrete substrate preferences (Table 1). The two residues to the amino side of the P1 aspartate (termed P2 and P3) have a limited effect on the substrate cleavage. In contrast, the P4 residue (three amino acid residues upstream from the P1 aspartate) appears to account for three distinct groups of substrate specificity (Table 1; ref. *9*).

Second, further classification was prompted by the functional differences between the caspases and is largely inherent to the nature of the N-terminal prodomain (reviewed ref. in *11*). Large prodomains contain various protein recruitment motifs such as DED and CARD *(12,13);* therefore, caspases bearing such prodomains, 2, 8, 9, and 10, are believed to be recruited upon apoptosis signaling, and to act as initiators. Other caspases, 3, 6, and 7, have shorter prodomains, are downstream targets of initiator caspases, and are considered as executioner or effector caspases. A last class of caspases, 1, 4, 5, 11, 12, 13, and 14, are labeled as "cytokine-processing" caspases. Only caspases 1 and 11 have been directly involved in cytokine activation (e.g., interleukin-1β) *(14,15).* In contrast, the remaining caspases are not as well characterized and are included in this category solely based on their homology to caspase-1 (Table 1). Despite its involvement in cytokine processing, it is important to remember that caspase-1 was first identified as a ced-3 homolog capable of inducing apoptosis *(16).*

3. Substrates of Caspases

Apoptosis is an orderly process that involves the specific cleavage of a definite set of physiological substrates by caspases. Over 40 substrates have been identified so far (reviewed in refs. *17–19*). Caspases do not merely degrade their targets, they also activate proteins that play an active role in the process of apoptosis. Caspases inactivate homeostatic proteins, turn antiapoptotic proteins into proapoptotic amplifiers of cell death, cleave structural proteins, and activate signaling pathways and effectors of the apoptotic phenotype *(18,19).* Exhaustive lists of caspase substrates have been presented elsewhere *(19);* Table 2 describes a functional classification of substrate cleavage. The involvement of caspases in apoptosis is a three-step process. Caspases first cleave substrates that further amplify the apoptotic signal. Caspases then dismantle proteins that ensure cell function and survival. Finally, they activate the effectors of the apoptotic phenotype, that is, enzymes that will package the dead cell for phagocytosis (Table 2 and references therein). Since the goal of this chapter is to focus on caspase involvement in neuronal cell death, examples of caspase substrates related to neurodegenerative diseases are included in Table 2.

Although the primary function of caspases is to hydrolyze peptide bonds in natural substrates, most assays designed to discover, characterize, and study regulation of these proteases rely on synthetic peptide substrates. The observation that caspases are active against tetrapeptides blocked at their N and C termini *(20)* permitted the development of synthetic substrates and inhibitors. The

Table 2
Caspase Cleavage of Cellular Substrates During Apoptosis is a Well-Planned and Orderly Process

Substrate class	Example	Caspase	References
Amplification of the apoptotic loop:			
Caspases (zymogens)	Auto- or caspase activation		Reviewed in (17)
Pro- and antiapoptotic proteins	Bcl-2	3, 9	(52,53)
	Bid	8	(54)
Homeostatic proteins			
Cellular repair mechanisms	DNA-protein kinase catalytic subunit	3	(55)
	PARP	3, 6, 7, 9	(56–59)
	p21	3, 7	(60,61)
Disruption of macromolecular synthesis	HnRNP proteins	3	(62)
	U1-70 kDa	3	(63)
Survival and cell cycle signals	Ras GTPase activating protein	3	(64)
	Rb	3	(65)
	CDC27	3	(66)
Induction of the apoptotic phenotype			
Cell structure demolition	Actin	1, 3	(67,68)
	NuMa	3, 6	(41)
	Lamins	6, 3	(37,38)
	ICAD/DFF	3	(69,70)
Cell structure remodeling	Fodrin/α_2-spectrin	3	(71)
(membrane blebbing)	Gelsolin	3	(72)
	PAK2	3	(73)
	ROCK I	3	(74,75)
Neurodegenerative diseases proteins	APP	3, 6, 7, 8	(30,45–48)
	Presenilins	1, 3, 6, 7, 8	(76)
	Tau	3, 6	(30,77)
	Huntingtin	3, 7	(78)

Table 3
Synthetic Caspase Substrates

Substrate	Caspase
VAD	All caspases
YVAD	1; 4; 5
LEHD	1; 2; 4; 5; 8; 9; 10
WEHD	1; 4; 5
VDVAD	2
DEVD	3; 6; 7; 8
VEID	6; 8
IETD	8; 9; 10

Adapted from refs. *3, 10,* and *50.*

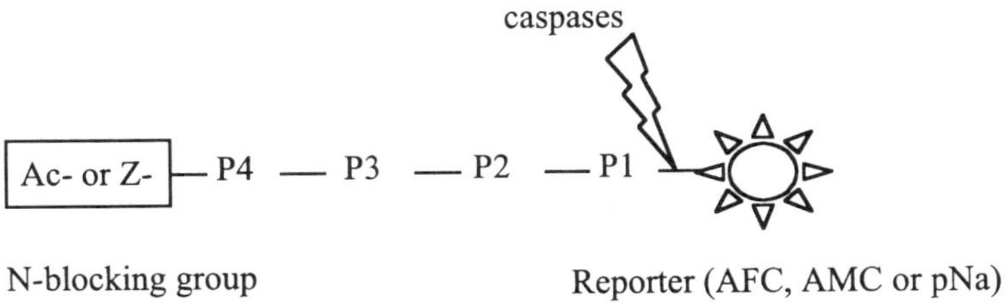

caspases

Ac- or Z- — P4 — P3 — P2 — P1

N-blocking group   Reporter (AFC, AMC or pNa)

Fig. 1. General schematic model of caspases tetrapeptide substrate. The peptide chain of a substrate is numbered sequentially with the "P" designation from the aspartic residue P1. The reporter is a colorimetric or fluorescent compound linked by the scissile bond to the C terminus of the substrate. The peptide side chains (P1–P4) are responsible for the specificity degree of each caspase. The N-blocking group is added to increase the substrate stability by preventing aminopeptidase digestion and to increase caspase catalysis, since many endopeptidase do not catalyze as readily when a free α-amine is present.

design of efficiently recognized peptidic substrates was based on known natural substrate sequences and combinatorial approaches with positional scanning of synthetic tetrapeptides (Table 3) *(3,10)*. Artificial substrates consist in an appropriate four amino acid recognition sequence linked via carboxy-terminal aspartate to a chromogenic or fluorogenic amine (Fig. 1) *(3)*. N-Blocking groups are either acetyl- (Ac-) or benzocarbonyl (Z-). The most commonly used reporter groups are *p*-nitroanilide (*p*Na, colorimetric detection by absorbance at 405–410 nm), 7-amino-4-trifluoromethyl-coumarin (AFC, fluorometric detection by excitation at 400 nm and emission at 480–520 nm; or colorimetric detection by absorbance at 380 nm), 7-amino-4-methylcoumarin (AMC, fluorometric detection by excitation at 365–380 nm and emission at 420–460 nm). The colorimetric assays are approximately 100-fold less sensitive than the fluorometric assays. Therefore, fluorometric assays are particularly useful in assaying low caspase concentrations and when investigating natural caspase inhibitors (*see* Subheadings 6. and 8.).

4. Inhibitors of Caspases

Anomalous activation of caspases can be detrimental; therefore protection mechanisms were evolved to avoid this problem. Inhibitors of caspases were first described in viruses. CrmA and p35 are two viral proteins that directly inhibit some caspases. CrmA, a cowpox virus serpin, inhibits most caspases but not group II caspases and caspase-6 (reviewed in ref. *21*). The baculoviral p35 is less specific and inhibits most caspases. The discovery of the baculoviral inhibitor of apoptosis (IAP) has prompted many researchers to investigate the existence of mammalian homologues. To date, more than half a dozen mammalian IAPs have been identified. Members of the IAP family include, X-linked IAP (X-IAP), c-IAP-1, c-IAP-2, and neuronal apoptosis inhibitory protein (NAIP) *(22)*. Other members are Livin, Survivin, and BRUCE *(23–25)*. Chau and colleagues *(26)* have recently identified Aven, a novel protein that prevents caspase activation by binding to Bcl-X_L and Apaf-1. Caspase inhibitors are of tremendous therapeutic value and have spurred an enormous amount of research toward the development of new therapeutic approaches.

A wide variety of synthetic peptide inhibitors of caspases have been developed to evaluate the implication of caspases in apoptotic processes. In contrast to fluorogenic substrates, caspase inhibitors allow studies to be performed in vitro, and in live animals and cells. These inhibitors are designed on the same principles used to generate caspase substrates as they rely on "caspase-specific" peptide recognition sequences *(3,10)*. These peptides act therefore as competitive inhibitors of caspases. They range from very broad-spectrum inhibitors consisting of one aspartate residue (e.g., Boc-D-fmk) to more sophisticated tetramers or even pentamers (e.g., Ac-VDVAD-CHO).

The mechanism of inhibition by synthetic peptides depends on the chemical groups conjugated to the peptides. Aldehyde-conjugated (–CHO) inhibitors do not form covalent bonds with the enzymes and are therefore reversible inhibitors. On the other hand, methylketone-based (-chloromethylketone, -cmk; or -fluoromethylketone, -fmk) peptides are irreversible inactivators of caspases since a thyomethylketone bond forms with the cysteine on the active site of the enzymes *(27)*. The methylketone groups are very reactive and consequently, these inhibitors exhibit decreased selectivity for caspases *(28)*, and are not exclusively specific for caspases. Schotte et al. *(29)* have demonstrated that z-DEVD-fmk can efficiently inhibit cathepsins at concentrations used to demonstrate the involvement of caspases. However, -fmk and -cmk conjugated peptides are much more permeable to the cells and are therefore more useful than their poorly permeable –CHO counterparts when studies are conducted in live cells.

The use of synthetic inhibitors specific to a given group of caspases can help pinpoint the caspase(s) involved in a given apoptotic model (described under Subheading 5.). However, although YVAD- and DEVD-based inhibitors are generally considered to be caspase-1- and caspase-3-specific, respectively, there are multiple caveats surrounding this issue. For example, Ac-DEVD-CHO is a very potent inhibitor of caspase-3, with a $K_i = 0.23$ nM; it is nevertheless quite potent against caspases-8 (0.92 nM) and -1 (15–18 nM) *(9,28)*. In one study, z-DEVD-fmk potently inhibited neuronal loss and apoptosis-mediated amyloid β-peptide production, which would theoretically indicate the involvement of caspase 3-like caspases. However, further analysis demonstrated that caspase-6 was in fact the culprit *(30)*. With this in mind, any experimental design involving caspase inhibitors should use multiple inhibitors as well as confirmation of caspase activation by other methods to avoid erroneous identification of the caspases involved in apoptosis (*see* Table 1 and Subheading 7.).

5. Detection of Caspase Involvement in Neuronal Apoptosis

The first approach to study the involvement of caspases in neuronal apoptosis involves the use of synthetic caspase inhibitors. This is a rather inexpensive, straightforward, and efficient method. It also does not require any sophisticated materials such as a fluorometer. In addition, caspase inhibitors can be used in intact living cells, avoiding artifacts that can be generated in lysed cells.

5.1. Materials and Solutions

Prepare stock solutions of the inhibitors at 10 mM in dry DMSO and store in small 10-µL aliquots at −20°C. The inhibitor is stable for over 3 mo at −20°C and at least 3 d at room temperature. To avoid contamination with moisture, bring frozen stock solutions to room temperature before opening. Aliquots can be thawed out without a significant loss of activity. The staurosporine stock (Sigma) is prepared in water at 10mM and stored at −20°C. Phosphate–buffered saline (PBS) pH 7.4 is prepared as follows: 0.058 M Na_2HPO_4; 0.017 M $NaH_2PO_4 \cdot H2O$; 0.068 M NaCl, and autoclaved 15 min at 121°C.

5.2. Methodology

To ensure optimal specificity, caspase inhibitors should not be used at concentrations higher than 1–5 µM.

5.2.1. Determination of the Involvement
of Caspases in Neuronal Cell Death

First, an experiment should be carried out with a broad-spectrum caspase inhibitor to test whether caspases are involved in the cell death pathway under study. Here, we will use staurosporine-induced cell death as an example.

1. Plate neurons in a 24-well plate on cover slips. We plate 50 µL per cover slip at 3×10^6 cells/mL as described by LeBlanc *(31)*. This may vary depending on the type of culture.
2. Allow neurons to attach and extend neurites. We perform all our experiments starting at 10 d in culture.
3. Pretreat neurons with 0.5 mL medium containing serum and 5 µM z-VAD-fmk or Boc-D-fmk. Three wells are pretreated with the inhibitors, while four are pretreated with medium containing serum and dimethylsulfoxide (DMSO) instead of the caspase inhibitors.
4. Incubate 1 h at 37°C/5% CO_2.
5. Add staurosporine to a final concentration of 10 µM to all but one well. This well will contain neither staurosporine nor z-VAD-fmk and is used as a negative control.
6. Incubate for the desired amount of time. Perform a time study using several time points. We generally incubate our primary human neurons for 24, 48, and 96 h, given their slow cell death. Shorter incubation times may be required for more vulnerable cells.

7. For long incubation times (>48 h), change medium and inhibitors every 2 d.
8. Fix cells in freshly prepared 4% paraformaldehyde/4% sucrose for 20 min at room temperature and permeabilize *(32)*.
9. Wash the cover slips 3 times for 5 min in PBS.
10. Stain for DNA fragmentation using TUNEL or Hoechst 33342 dye. Hoechst 33342 is less labor intensive and less expensive than TUNEL. However, TUNEL allows easier detection of apoptosis, since only dead cells stain, whereas Hoechst 33342 stains both live and dead cells. For Hoechst 33342 staining, proceed as follows.
11. Incubate cells in PBS containing 0.4 µg/mL Hoechst 33342 for 20 min at room temperature.
12. Rinse the cover slips twice in PBS.
13. Mount on glass slides using a photobleaching-resistant mounting media (e.g., Geltol, Shandon, PA).
14. Assess the extent of cell death by measuring the number of fragmented and shrunken nuclei under a fluorescence microscope. Count 400–500 cells per condition. If caspases are involved, the samples incubated in the presence of the caspase inhibitor should display a significant reduction in the extent of cell death.
15. Apoptosis (and inhibition of apoptosis) can be confirmed using various methods. Cells can be collected prior to fixing and the DNA extracted. The DNA is separated on 1% agarose gels and stained with ethidium bromide. Apoptosis is characterized by a typical DNA ladder pattern (for a detailed method, *see* ref. *33*). Alternatively, Annexin V staining can be used to monitor early plasma membrane changes during apoptosis *(34)*.

5.2.2. Identification of the Specific Groups of Caspases Involved in Apoptosis

Once the involvement of caspases is ascertained, a more detailed study can be performed using specific inhibitors. Table 4 provides a guide for the selection of the appropriate inhibitors. Following is a typical experiment.

1. Plate cells on coverslips as indicated above. Prepare one plate per time point.
2. z-WEHD-fmk, z-DEVD-fmk, and z-VEID-fmk will be used as inhibitors at 0, 0.5, 1, and a concentration of 5 µ*M*. Staurosporine (STS) is used at a final concentration of 10 µ*M* to induce apoptotis.

Table 4
Inhibition Efficiency of Commonly Used Inhibitors Against Caspases

				Inhibited by			
Enzyme	YVAD	WEHD	DEVD	VDVAD	VEID	IETD	LEHD
Caspase 1	++	+++	+	ND	++	++	ND
Caspase 4	+	++	++	ND	ND	+	ND
Caspase 5	+	++	+	ND	ND	+	ND
Caspase 3	−	−	++	ND	+	+	ND
Caspase 7	−	−	++	ND	−	−	ND
Caspase 2	−	−	−	++	−	−	−
Caspase 6	−	−	++	ND	+++	++	−
Caspase 8	+/−	++	++	ND	++	++	ND
Caspase 9	−	+/−	+	ND	ND	+	++
Caspase 10	+/−	+/−	++	ND	++	++	ND

−, no inhibition; +, slight inhibition; ++, strong inhibition; +++, very strong inhibition; ND, not determined.
Compiled from refs. *9, 10,* and *28.*

3. Table 5 represents the organization of the plates.
4. Based on your first experiment, choose a time where the insult induces significant cell death.
5. Pretreat the cells for 1 h with the appropriate inhibitor at the start of each experiment prior to the induction of apoptosis.
6. Fix the cells at each timepoint in 4% paraformaldehyde/4% sucrose for 20 min at room temperature *(32).*
7. Wash cells 3 times for 5 min in PBS.
8. Incubate cells in PBS containing 0.4-µg/mL Hoechst 33342 for 20 min at room temperature.
9. Rinse the cover slips twice in PBS.
10. Mount on glass slides using a photobleaching-resistant mounting media (e.g., Geltol, Shandon, PA).
11. Count cell death.

The various caspase inhibitors will display different inhibition profiles indicative of the group of caspases involved in cell death.

5.3. Points to Remember

1. Caspase inhibitors will only help you pinpoint a *group* of caspases, not one particular caspase.
2. To determine the exact caspase(s) involved, *see* Subheadings 6. and 7.

Table 5
Schematic Representation of a Typical
24-Well Plate Experimental Design Using Caspase Inhibitors[a]

[Caspase inhibitor] (µM)	10 µM staurosporine					
	−	+	−	+	−	+
0	WEHD	WEHD	DEVD	DEVD	VEID	VEID
0.5	WEHD	WEHD	DEVD	DEVD	VEID	VEID
1	WEHD	WEHD	DEVD	DEVD	VEID	VEID
5	WEHD	WEHD	DEVD	DEVD	VEID	VEID

[a]One 24-well plate per time point.

6. Fluorimetric Assays for the Quantitative In Vitro Determination of Caspase Activity

The second approach to study the involvement of caspases in neuronal apoptosis involves the use of synthetic caspase substrates. This assay is highly sensitive, nonradioactive, convenient, and very rapid, as the entire protocol can be performed in less than 2 h. It is useful to analyze caspase activation and to determine a certain specific stage of the apoptotic process. Alternative methods to determine caspase activity, based on the cleavage of natural substrates, require purified proteins or radiolabeled, in vitro-transcribed/translated proteins and are therefore more labor intensive.

A two-step strategy using fluorogenic assays can be used to demonstrate caspase activation during neuronal cell death. First, an increase in specific activity for each synthetic substrate (DEVD, VEID, YVAD, etc.) can be used to demonstrate caspase activation and the group of caspases involved. Second, kinetic studies of enzyme cleavage activity (e.g., DEVDase activity) using different aldehyde-conjugated caspase inhibitors can be performed to determine the sensitivity to these inhibitors and subsequently the specific caspase(s) activated. We describe here a method we use in our laboratory to assess endogenous caspase-6 activation during neuronal cell death. Caspase-6 (Mch2α) is a key executioner of apoptosis in human neurons *(35)*, responsible either partially or totally for the proteolytic cleavage of many cellular substrates. Many of these substrates are nuclear proteins: topoisomerase I *(36)*, lamin A and B *(37, 38)*, lamin B receptor *(39)*, PARP *(40)*, and Numa *(41)*. Caspase-6 also cleaves cytoskeletal proteins such as keratin and β-catenin *(42, 43)*, cytoplasmic caspase-3 *(44)*, and plasma membrane proteins, amyloid precursor protein (APP) *(30,45–48)*, and focal adhesion kinase (FAK) *(46)*. Caspase-6 protease activity can be used to monitor neuronal apoptosis. One way to measure caspase-6 activity is based on a fluorescent substrate, Ac-VEID-AFC (acetyl-Val-Glu-Asp-7-amino-4-trifluoromethyl coumarin), which mimics the known cleavage site of lamin A, for which caspase-6 shows the highest affinity relative to other caspases *(3,49)*. The AFC conjugate normally emits blue light (max = 400 nm), but upon proteolytic cleavage by caspase-6 or closely related caspases, the free AFC emits a yellow-green fluorescence at 505 nm. Comparison of the fluorescence emission of an apoptotic sample with an uninduced control allows one to determine the fold-increase in protease activity.

6.1. Materials and Solutions

1. The fluorescent products can typically be quantitated using a fluorometer or a fluorescence microplate reader, which facilitates high-throughput analysis and requires relatively small assay volumes. We use in our laboratory a Fluoromark™ microplate fluorometer (Bio-Rad). As mentioned above, each fluorogenic reporter group has different excitation and emission wavelength, so the fluorometer must be equipped with specific filters (i.e., 390-nm excitation filter and a 535-nm emission filter) for experiments using AFC.
2. 96-well flat-bottom polystyrene plates with opaque walls to prevent well-to-well crosstalk and optically clear bottoms should be used. Bottoms should be thinner than conventional polystyrene plates, resulting in lower background fluorescence and enabling readings down to 340 nm. We currently use the Falcon Microtest 96-well assay plate (Optilux) from Becton Dickinson.
3. Prepare stock solution of caspase-6-like enzyme substrate Ac-VEID-AFC in dimethyl sulfoxide (DMSO) at 5 mg/mL. The stock solution is stable for at least 1 yr at −20°C. Hydrolysis of the substrate is revealed by the appearance of a yellow color. Table 3 shows selective substrates for each caspase. A number of these different synthetic tetrapeptide substrates are commercially available from various suppliers. We routinely use Ac-DEVD-AFC to assay caspase-3- and -7-like activity and Ac-IETD-AFC to assay caspase-8-like activity (Table 3). All sub-

strates used in our laboratory are purchased from Biomol Research Laboratories.

4. Prepare stock solution of AFC 200 μ*M* in dry DMSO. The stock solution is also stable for at least 1 yr at –20°C.

5. Cell lysis buffer without protease inhibitors: 50 m*M* 4-(2-hydroxyethyl)-1-piperazineethanesulfonic acid (HEPES) pH 7.4, 0.1% (w/v) 3-[(3-cholamidopropyl)-dimethyl-ammonio]-1-propanesulfonate (CHAPS), 1 mM dithiothreitol (DTT), 0.1 m*M* EDTA (Biomol Research Laboratories).

6. 2.5X Caspase reaction buffer: 50 m*M* (piperazine-N-N'-bis (2-ethanesulfonic acid (PIPES), 75 m*M* NaCl, 25 m*M* DTT, 2.5 m*M* EDTA, 0.25% (w/v) CHAPS, 25% (w/v) sucrose, pH: 7.2 *(50)*. Do not add DTT to the 2.5X caspase reaction buffer stock, add 25 m*M* DTT immediately before use to the aliquotted 2.5X caspase buffer and caspase reaction buffer.

7. Phosphate-buffered saline (PBS), pH 7.4: 0.058 *M* Na_2HPO_4; 0.017 *M* $NaH_2PO_4 \cdot H_2O$; 0.068 *M* NaCl.

8. Recombinant caspase-6 (R-Csp-6) (BD Pharmingen). The enzyme should be stored in small aliquots at –80°C, and kept on ice throughout the experiment. This enzyme loses activity after a few cycles of freezing/thawing. Dilute the R-Csp-6 in cell lysis buffer to a final concentration of 1 ng/μL just before use.

6.2. Methodology

1. 6×10^6 neurons are treated with the apoptotic insult in 6-well culture plates at 37°C/5% CO_2. For comparative analysis, include nontreated cells. Use at least 2×10^6 neurons/well. Using too few cells may be below the detection limit for caspase activity.

2. All subsequent work should be done at 4°C or on ice.

3. For time-course studies, collect the cells at specific times by washing once with cold PBS.

4. Add 100 μL of chilled cell lysis buffer, tilt the plate gently, and scrape the cells into the buffer using a plastic scraper or rubber policeman. Transfer the cells and buffer to labeled, prechilled 1.5-mL microcentrifuge tubes. The amount of cell lysis buffer to be added to the cells is determined by the number of cells present (this can be estimated from the number of cells initially cultured). We add 100 μL of cold cell lysis buffer per 6×10^6 cells cultured in a 6-well plate and 600 μL for 36×10^6 cells in a 75-cm² culture flask.

5. Incubate on ice for 10 min. Pellet insoluble material by centrifugation for 10 min at 16,000g at 4°C. The clear supernatant can be used immediately or stored at −80°C.

6. This should yield a cell lysate with an approximate protein concentration of 1 mg/mL. The protein content of the cell lysate can be measured using a protein determination assay that is compatible with detergents present in the cell lysis buffer, e.g., BCA Protein Assay (Pierce Chemicals) or Bradford protein assay (Bio-Rad).

7. Using the fluorometer software, define microplate sample positions (position of standards, blanks, unknowns, empty cells, different concentrations, etc.) and test protocols (measurements timing, filters for excitation and emission, gain, etc.). When defining the test procedure in the fluorometer software, it is important to adjust the gain function that provides optimal sensitivity for the fluorometer. For example, determine the optimal gain value at the highest AFC standard concentration to obtain the maximum sensitivity of detection.

8. Preheat the instrument at 37°C.

9. The caspase-6 activity is measured at 37°C every 2 min for 1 h to determine the linear range of activity. If needed, when the activity is low, extend measures every 10 min for 2 h. The rate of hydrolysis is determined from the observed progress curves. It is important to ensure that the rates are calculated from the linear portion of the progress curve, and single-time-point assay should be avoided. The Fluoromark software we use in our laboratory displays a real-time graphics screen of the defined microplate format, which is helpful to monitor the progress of the reaction (Fig. 2).

10. We recommend that assays be performed in small volumes (50–100 µL).

11. Prepare dilutions of AFC standards in 1X caspase reaction buffer. The linear range of the assay is 0.4 to 50 µM. Additional controls that should be included in this assay are (Fig. 2):
    a. Blank: reactions where no cell lysate and no Ac-VEID-AFC are added to 1X caspase reaction buffer.
    b. Ac-VEID-AFC alone: reactions where no cell lysate is added to the 1X caspase reaction buffer.
    c. A control with a caspase inhibitor should be used to demonstrate specific caspase activation.
    d. A control using 5 ng of R-Csp-6 should be used as a positive control.

A

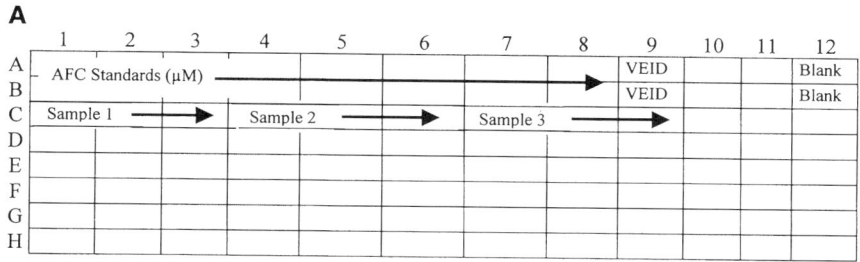

B

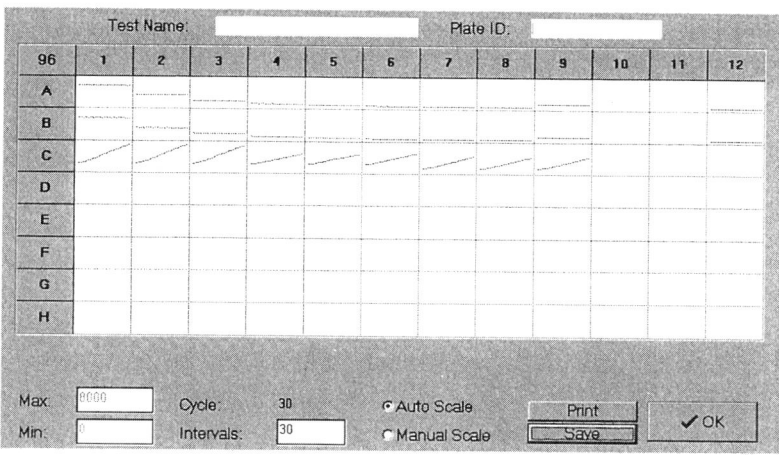

Fig. 2. Example of a typical caspase assay design. (**A**) Schematic representation of a typical 96-well plate experimental design for caspase assay. A1-8 and B1-8, serial dilutions of AFC standards; A9 and B9, control with substrate (Ac-VEID-AFC) alone where no cell lysate is added; samples 1–3, different conditions tested, i.e., drug 1, drug 2, and nontreated control, respectively. (**B**) Final results obtained from the designated experiment as shown on the real-time graphics screen (Fluoromark software, Bio-Rad). The software displays this graphics screen in a defined microplate format. Caspase activity is measured at 37°C every 2 min for 1 h.

12. We usually prepare a master mixture of the reaction mixture containing: 20 μL 2.5X caspase buffer, 0.5 μL substrate Ac-VEID-AFC (5-mg/mL stock solution to make a final concentration of 68.5 μ*M*), and 19.5 μL H$_2$O for each well, multiplied by the number of wells tested. We then add 40 μL from the master mixture to each well.

13. Thereafter, rapidly add the cell lysate to the appropriate wells. We typically use 5 µg of total proteins in a 10-µL volume. The total reaction volume must be kept constant (50 µL) and therefore cell lysis buffer can be used to replace the volume normally occupied by the cell lysate.
14. Standards should be done in duplicate and samples in triplicate (Fig. 2).
15. Insert the plate and immediately read the sample's fluorescence (timing was set in step 9).
16. The data from the fluorometer can be viewed in a spreadsheet program. The fluorometer assigns a fluorescence value for each sample at every time point. To calculate the caspase specific activity, one must select a time point within the linear range of activity of all samples. Based on the AFC standard curve, the amount of released AFC is measured and the specific activity of the caspase determined as nmoles of released AFC per microgram total protein/per minute.
17. The readings from the background controls (reactions where no cell lysate is added or where no Ac-VEID-AFC substrate is added) must be subtracted from the experimental results prior to calculating the specific activity. Untreated neurons can show basal caspase activity. Therefore it is important to show the increase of caspase activity in time. This assay is suitable for systems where only a determination of the catalytic activity is required. Kinetic parameters can only be determined for pure caspases not complex, undefined mixtures.

6.3. Troubleshooting and Hints

The caspase activity assay protocol is straightforward and does not involve extensive optimization. However, if you do not see the expected results, check the following.

1. Check that the fluorometer has the right filters.
2. Check that the fluorometer is reading at the right wavelength.
3. Check that nonspecific protease cleavage of specific substrates may be detected. Supplement the caspase buffer with protease inhibitors. We use in our laboratory 0.5-µg/mL leupeptin, 0.1-µg/mL pepstatin, and 1-µg/mL N^{α}–p-tosyl-L-lysine chloromethyl ketone (TLCK) as protease inhibitors. Furthermore, a new package of tubes should be used and working with new gloves is recommended to avoid protease contamination of the samples.

The use of broad-spectrum protease inhibitors or cysteine protease inhibitors such as α_2-macroglobulin, iodoacetamide, and N-ethylmaleimide should be avoided.

4. Caspases need to be reduced in order to retain full activity. Therefore they require high concentrations of reducing agents such as DTT to achieve maximal activity. Buffers containing DTT should be freshly prepared.

5. Data should be analyzed with great care due to the important overlap in the substrate preferences of the caspases. Although a particular tetrapeptide substrate may be referred to as a "caspase-6 substrate" or a "caspase-3 substrate," this should not be taken to imply that it is a substrate that can only be cleaved by that particular caspase but rather that the substrate is cleaved efficiently by that caspase (Table 3).

6. Essentially the same procedure can be performed with a chromogenic instead of a fluorogenic tetrapeptide substrate. Although this technique is slightly less sensitive and requires more cells, it does not require a fluorometer, but it can be performed with a normal spectrophotometer.

7. Immunological Detection of Caspase Activation

The use of caspase inhibitors and fluorometric substrates allows the identification of groups of caspases rather than a specific caspase. To determine the implicated caspase(s), Western blotting can be used against caspases of the identified group. Caspase activation can be monitored either by the disappearance of the proenzyme or by the appearance of the active p20 or p10 fragments (Table 1).

Caspase-specific antibodies are widely distributed and are available from a number of commercial sources.

7.1. Methodology

1. Neurons are either treated with the apoptotic insults or left untreated in 6-well plates or 75-cm³ flasks.

2. After incubation, extract proteins in 200 µL/well (600 µL for the flasks) of NP-40 lysis buffer (50 mM Tris HCl, pH 8.0, 150 mM NaCl, 1% NP-40, 5 mM EDTA, 0.05% PMSF, 0.1-µg/mL pepstatin A, 1 mg/mL TLCK, 0.5-µg/mL leupeptin) *(51)*.

3. Incubate cell lysates 10 min on ice.

4. Microfuge 5 min.

5. Transfer supernatants, i.e., NP-40 soluble fraction (NP-SF), to new tubes.
6. Add 50 µL of 0.1% SDS to the pellets.
7. Boil 3 min. Microfuge 5 min.
8. Transfer supernatants, i.e., NP-40–insoluble fraction (NP-IF), to new tubes.
9. Determine the protein content of each sample using standard methods.
10. Separate 40–100 µg protein/lane on 15% SDS-PAGE. Include both NP-SF and NP-IF.
11. Transfer proteins to a PVDF membrane and blot according to standard protocols. The blots are to be probed using antibodies directed against each of the caspases in the suspected group (I, II, or III; *see* Table 1).

7.2. Hints and Troubleshooting

The method described above is straightforward and yet very powerful. However, there are a few limitations:

1. The number of cells available can be a limiting factor rendering the detection of the active fragments more difficult.
2. In the case of heterogeneous cultures or tissues, erroneous conclusions may be drawn. For example, in a whole-brain extract, caspase-3 activation can be detected; however, this activation may be confined to a specific cell type (e.g., astrocytes). To overcome these problems, immunocytochemistry is the approach of choice (*see* the chapter by Kevin Roth).
3. Further assessment of the involvement of a given caspase in the cell death pathway under study can be done by direct microinjection of active recombinant caspases into live cells as described in the chapter by Zhang and LeBlanc.

8. Assaying for Endogenous Caspase Inhibitors in Neurons

We describe here a method we use in our laboratory to assess the direct effect of neuroprotective drugs on caspase-mediated cell death and to screen neuronal extracts for natural caspase-6 inhibitors using synthetic substrates. This protocol is adapted from the caspase activity assay to determine specific endogenous caspases activation during cell death described under Subheading 6.

8.1. Materials and Solutions

All materials and solutions are as described under Subheading 6.1.

8.2. Methodology

1. Treat 6×10^6 neurons/well for each sample in a 6-well culture plate with the apoptotic insult in absence or presence of different concentrations of the known neuroprotective drug. For comparative analysis, include nontreated cells.
2. Follow the procedure as described under Subheading 6.2., steps 2–9.
3. Prepare dilutions of AFC standards in 1X caspase reaction buffer. The linear range of the assay is 0.4 to 50 μM. Additional controls that should be included in this assay are as follows.
   a. Blank: reactions where no cell lysate, no Ac-VEID-AFC and no enzyme were added to the 1X caspase reaction buffer.
   b. Control Ac-VEID-AFC alone: reactions where no cell lysate and no enzyme were added to the 1X caspase reaction buffer.
   c. Control recombinant active caspase-6 (R-Csp-6) alone: reaction where no cell lysate is added to the 1X caspase reaction buffer.
   d. Standards should be done in duplicate and samples in triplicate as shown in Fig. 2.
4. Prepare a master mixture of the reaction mixture containing: 20 μL 2.5X caspase buffer, 0.5 μL substrate (Ac-VEID-AFC), and 14.5 μL H_2O for each well, multiplied by the number of wells tested. Then add 35 μL from the reaction mixture to each well.
5. Thereafter, rapidly add cell lysate to the appropriate wells. We typically use 5 μg of total proteins in a 10-μL volume. The total reaction volume (50 μL) must be kept constant and therefore caspase lysis buffer can be used to replace the volume normally occupied by the cell lysate.
6. Add 5 μL (5 ng) of purified R-Csp-6 (BD Pharmingen) to each sample. The enzyme may be activated in caspase buffer 15 min at 37°C *(50)*. However, in our experience there is no need to activate the enzyme as the activity is measured at 37°C every 2 min for at least 1 h to determine the linear range of activity.
7. Read the caspase activity immediately after addition of the R-Csp-6.
8. Results should be expressed and analyzed as described under Subheading 6.2.

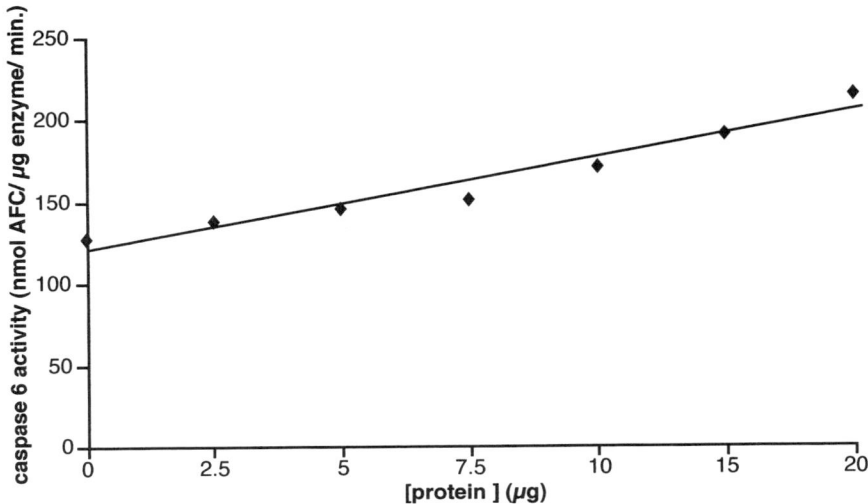

Fig. 3. Effect of protein concentration on recombinant caspase-6 activity. The R-Csp-6 activity is measured at 37°C every 2 min for 1 h in the presence of various concentrations of bovine serum albumin (BSA fraction V, Sigma). Protein concentration was determined by Bradford assay. Proteins (2.5–20 µg) were added to 5 ng of R-Csp-6 (Pharmingen) in caspase assay buffer and 68.5 µM Ac-VEID-AFC. Based on AFC standards curve, the amount of released AFC was measured and the specific activity of the caspase determined as nmol released AFC/µg enzyme/min.

8.3. Hints and Troubleshooting

Caspase-6 activity increases with increasing protein concentrations in the assay mixture (Fig. 3). Therefore, caution must be exercised in the calculations of enzyme specific activity and to assess the effect of neuroprotective drugs on R-Csp-6 activity. Bovine serum albumin (BSA) is always added to control for total protein in the assay. This may not be true for other caspases.

9. Conclusion

The major feature distinguishing the caspases from one another is the P1 substrate preference (3,10). However, the synthetic substrates and inhibitors used in most studies display a tremendous lack of specificity. In addition, the catalytic efficiency, concentration, and cellular distribution of caspases hinder the interpretation of data. In other words, at any given time, we may only be observing the most active/most concentrated caspase. To circumvent this problem and to determine which caspases are involved, a combination of different substrates and inhibitors should be used at various times of apoptosis. Moreover, various methods should be used to confirm the presence of any suspected active caspase.

References

1. Johnson, E. M. J., Chang, J. Y., Koike, T., and Martin, D. P. (1989) Why do neurons die when deprived of trophic factor? *Neurobiol. Aging* **10,** 549–552.
2. Thornberry, N. A. and Lazebnik, Y. (1998) Caspases: enemies within. *Science* **281,** 1312–1316.
3. Thornberry, N. A., Rano, T. A., Peterson, E. P., et al. (1997) A combinatorial approach defines specificities of members of the caspase family and granzyme B. Functional relationships established for key mediators of apoptosis. *J. Biol. Chem.* **272,** 17,907–17,911.
4. Uren, A. G., O'Rourke, K., Aravind, L., et al. (2000) Identification of paracaspases and metacaspases: two ancient families of caspase-like proteins, one of which plays a key role in MALT lymphoma. *Mol. Cell* **6,** 961–967.
5. Mittl, P. R., Di Marco, S., Krebs, J. F., et al. (1997) Structure of recombinant human CPP32 in complex with the tetrapeptide acetyl-asp-val-ala-asp fluromethyl ketone. *J. Biol. Chem.* **272,** 6539–6547.
6. Rotonda, J., Nicholson, D. W., Fazil, K. M., et al. (1996) The three-dimensional structure of apopain/CPP32, a key mediator of apoptosis. *Nat. Struct. Biol.* **3,** 619–625.
7. Walker, N. P., Talanian, R. V., Brady, K. D., et al. (1994) Crystal structure of the cysteine protease interleukin 1-beta converting enzyme: a $(p20/p10)_2$ homodimer. *Cell* **78,** 343–352.
8. Wilson, K. P., Black, J. A., Kim, E. E., et al. (1994) Structure and mechanism of interleukin-1 beta converting enzyme. *Nature* **370,** 270–275.
9. Garcia-Calvo, M., Peterson, E. P., Leiting, B., Ruel, R., Nicholson, D. W., and Thornberry, N. A. (1998) Inhibition of human caspases by peptide based and macromolecular inhibitors. *J. Biol. Chem.* **273,** 32,608–32,613.
10. Talanian, R. V., Quinlan, C., Trautz, S., et al. (1997) Substrate specificities of caspase family proteases. *J. Biol. Chem.* **272,** 9677–9682.
11. Wolf, B. B. and Green, D. R. (1999) Suicidal tendencies: apoptotic cell death by caspase family proteinases. *J. Biol. Chem.* **274,** 20,049–20,052.
12. Chinnaiyan, A. M., O'Rourke, K., Tewari, M., and Dixit, V. M. (1995) FADD, a novel death domain-containing protein, interacts with the death domain of Fas and initiates apoptosis. *Cell* **81,** 505–512.
13. Hofmann, K., Bucher, P., and Tschopp, J. (1997) The CARD domain: a new apoptotic signalling motif. *Trends Biochem. Sci.* **22,** 155–156.
14. Kuida, K., Lippke, J. A., Ku, G., et al. (1995) Altered cytokine export and apoptosis in mice deficient in interleukin-1 beta converting enzyme. *Science* **267,** 2000–2003.
15. Wang, S., Miura, M., Jung, Y. K., Zhu, H., Li, E., and Yuan, J. (1998) Murine caspase-11, an ICE-interacting protease, is essential for the activation of ICE. *Cell* **92,** 501–509.

16. Miura, M., Zhu, H., Rotello, R., Hartwieg, E. A., and Yuan, J. (1993) Induction of apoptosis in fibroblasts by IL-1 beta-converting enzyme, a mammalian homolog of the *C. elegans* cell death gene *ced-3*. *Cell* **75,** 653–660.
17. Cohen, G. M. (1997) Caspases: the executioners of apoptosis. *Biochem. J.* **326,** 1–16.
18. Cryns, V. and Yuan, J. (1998) Proteases to die for. *Genes Dev.* **12,** 1551–1570.
19. Earnshaw, W. C., Martins, L. M., and Kaufmann, S. H. (1999) Mammalian caspases: structure, activation, substrates, and functions during apoptosis. *Annu. Rev. Biochem.* **68,** 382–424.
20. Thornberry, N. A., Bull, H. G., Calaycay, J. R., et al. (1992) A novel heterodimeric cysteine protease is required for interleukin-1 beta processing in monocytes. *Nature* **356,** 768–774.
21. Ekert, P. G., Silke, J., and Vaux, D. L. (1999) Caspase inhibitors. *Cell Death Diff.* **6,** 1081–1086.
22. Goyal, L. (2001) Cell death inhibition: keeping caspases in check. *Cell* **104,** 805–808.
23. Hauser, H. P., Bardroff, M., Pyrowolakis, G., and Jentsch, S. (1998) A giant ubiquitin-conjugating enzyme related to IAP apoptosis inhibitors. *J. Cell. Biol.* **141,** 1415–1422.
24. Kasof, G. M. and Gomes, B. C. (2001) Livin, a novel inhibitor-of-apoptosis (IAP) family member. *J. Biol. Chem.* **276,** 3238–3246.
25. Li, F., Ackermann, E. J., Bennett, C. F., et al. (1999) Pleiotropic cell-division defects and apoptosis induced by interference with survivin function. *Nat. Cell Biol.* **1,** 461–466.
26. Chau, B. N., Cheng, E. H., Kerr, D. A., and Hardwick, J. M. (2000) Aven, a novel inhibitor of caspase activation, binds Bcl-xL and Apaf-1. *Mol. Cell* **5,** 31–40.
27. Thornberry, N. A., Peterson, E. P., Zhao, J. J., Howard, A. D., Griffin, P. R., and Chapman, K. T. (1994) Inactivation of interleukin-1 beta converting enzyme by peptide (acyloxy)methylketones. *Biochemistry* **33,** 3934–3940.
28. Margolin, N., Raybuck, S. A., Wilson, K. P., et al. (1997) Substrate and inhibitor specificity of interlukin-1 beta converting enzyme and related caspases. *J. Biol. Chem.* **272,** 7223–7228.
29. Schotte, P., Declercq, W., Van Huffel, S., Vandenabeele, P., and Beyaert, R. (1999) Non-specific effects of methylketone peptide inhibitors of caspases. *FEBS Lett.* **442,** 117–121.
30. LeBlanc, A., Liu H., Goodyer, C., Bergeron, C., and Hammond, J. (1999) Caspase-6 role in apoptosis of human neurons, amyloidogenesis, and Alzheimer's disease. *J. Biol. Chem.* **274,** 23,426–23,436.
31. LeBlanc, A. (1995) Increased production of 4 kDa amyloid beta peptide in serum deprived human primary neuron cultures: possible involvement of apoptosis. *J. Neurosci.* **15,** 7837–7846.
32. Harlow, E. and Lane, D. (1999) Staining cells, in *Using Antibodies: A Laboratory Manual* (Harlow, E. and Lane, D., eds.), Cold Spring Harbor Laboratory, Cold Spring Harbor, NY, pp. 123–125.
33. Moyse, E. and Michel, D. (1997) Analyses of apoptosis-associated DNA fragmentation, in *Apoptosis Techniques and Protocols* (Poirier, J., ed.), Humana Press, Totowa, NJ, pp. 133–160.

34. Koopman, G., Reutelingsperger, C. P., Kuijten, G. A., Keehnen, R. M., Pals, S. T., and van Oers, M. H. (1994) Annexin V for flow cytometric detection of phosphatidylserine expression on B cells undergoing apoptosis. *Blood* **84,** 1415–1420.

35. Zhang, Y., Goodyer, C., and LeBlanc, A. (2000) Selective and protracted apoptosis in human primary neurons microinjected with active caspase-3, -6, -7, and -8. *J. Neurosci.* **20,** 8384–8389.

36. Samejima, K., Svingen, P. A., Basi, G. S., et al. (1999) Caspase-mediated cleavage of DNA topoisomerase I at unconventional sites during apoptosis. *J. Biol. Chem.* **274,** 4335–4340.

37. Lazebnik, Y. A., Cole, S., Cooke, C. A., Nelson, W. G., and Earnshaw, W. C. (1993) Nuclear events of apoptosis in vitro in cell-free mitotic extracts: a model system for analysis of the active phase of apoptosis. *J. Cell Biol.* **123,** 7–22.

38. Rao, L., Perez, D., and White, E. (1996) Lamin proteolysis facilitates nuclear events during apoptosis. *J. Cell Biol.* **135,** 1441–155.

39. Duband-Goulet, I., Courvalin, J. C., and Buendia, B. (1998) LBR, a chromatin and lamin binding protein from the inner nuclear membrane, is proteolyzed at late stages of apoptosis. *J. Cell Sci.* **111,** 1441–1451.

40. Fernandes-Alnemri, T., Litwack, G., and Alnemri, E. S. (1995) Mch2, a new member of the apoptotic Ced-3/Ice cysteine protease gene family. *Cancer Res.* **55,** 2737–2742.

41. Hirata, H., Takahashi, A., Kobayashi, S., et al. (1998) Caspases are activated in a branched protease cascade and control distinct downstream processes in Fas-induced apoptosis. *J. Exp. Med.* **187,** 587–600.

42. Caulin, C., Salvesen, G. S., and Oshima, R. G. (1997) Caspase cleavage of keratin 18 and reorganization of intermediate filaments during epithelial cell apoptosis. *J. Cell Biol.* **138,** 1379–1394.

43. Van de Craen, M., Berx, G., Van den Brande, I., Fiers, W., Declercq, W., and Vandenabeele, P. (1999) Proteolytic cleavage of beta-catenin by caspases: an *in vitro* analysis. *FEBS Lett.* **458,** 167–170.

44. Xanthoudakis, S., Roy, S., Rasper, D., et al. (1999) Hsp60 accelerates the maturation of pro-caspase-3 by upstream activator proteases during apoptosis. *EMBO J.* **18,** 2049–2056.

45. Barnes, N. Y., Li, L., Yoshikawa, K., Schwartz, L. M., Oppenheim, R. W., and Milligan, C. E. (1998) Increased production of amyloid precursor protein provides a substrate for caspase-3 in dying motoneurons. *J. Neurosci.* **18,** 5869–5880.

46. Gervais, F. G., Xu, D., Robertson, G. S., et al. (1999) Involvement of caspases in proteolytic cleavage of Alzheimer's amyloid-beta precursor protein and amyloidogenic A beta peptide formation. *Cell* **97,** 395–406.

47. Pellegrini, L., Passer, B. J., Tabaton, M., Ganjei, J. K., and D'Adamio, L. (1999) Alternative, non-secretase processing of Alzheimer's beta-amyloid precursor protein during apoptosis by caspase-6 and -8. *J. Biol. Chem.* **274,** 21,011–21,016.

48. Weidemann, A., Paliga, K., D rrwang, U., Reinhard, F. B., Schuckert, O., Evin, G., and Masters, C. L. (1999) Proteolytic processing of the Alzheimer's disease amyloid precursor protein within its cytoplasmic domain by caspase-like proteases. *J. Biol. Chem.* **274,** 5823–5829.

49. Takahashi, A., Musy, P. Y., Martins, L. M., Poirier, G. G., Moyer, R. W., and Earnshaw, W. C. (1996) CrmA/SPI-2 inhibition of an endogenous ICE-related protease responsible for lamin A cleavage and apoptotic nuclear fragmentation. *J. Biol. Chem.* **271**, 32,487–32,490.

50. Stennicke, H. R. and Salvesen, G. S. (1997) Biochemical characteristics of caspases-3, -6, -7, and -8. *J. Biol. Chem.* **272**, 25,719–25,723.

51. Harlow, E. and Lane, D. (1999) Immunoprecipitation, in *Using Antibodies: A Laboratory Manual* (Harlow, E. and Lane, D., eds.), Cold Spring Harbor Laboratory, Cold Spring Harbor, NY, pp. 221–266.

52. Cheng, E. H., Kirsch, D. G., Clem, R. J., et al. (1997) Conversion of Bcl-2 to a Bax-like death effector by caspases. *Science* **278**, 1966–1968.

53. Tomicic, M. T. and Kaina, B. (2001) Hamster Bcl-2 protein is cleaved in vitro and in cells by caspase-9 and caspase-3. *Biochem. Biophys. Res. Commun.* **281**, 404–408.

54. Li, H., Zhu, H., Xu, C. J., and Yuan, J. (1998) Cleavage of BID by caspase 8 mediates the mitochondrial damage in the Fas pathway of apoptosis. *Cell* **94**, 491–501.

55. Casciola-Rosen, L. A., Anhalt, G. J., and Rosen, A. (1995) DNA-dependent protein kinase is one of a subset of autoantigens specifically cleaved early during apoptosis. *J. Exp. Med.* **182**, 1625–1634.

56. Alnemri, E. S., Fernandes-Alnemri, T., and Litwack, G. (1995) Cloning and expression of four novel isoforms of human interleukin-1 beta converting enzyme with different apoptotic activities. *J. Biol. Chem.* **270**, 4312–4317.

57. Duan, H., Orth, K., Chinnaiyan, A. M., et al. (1996) ICE-LAP6, a novel member of the ICE/Ced-3 gene family, is activated by the cytotoxic T cell protease granzyme B. *J. Biol. Chem.* **271**, 16,720–16,724.

58. Kaufmann, S. H., Desnoyers, S., Ottaviano, Y., Davidson, N. E., and Poirier, G. G. (1993) Specific proteolytic cleavage of poly(ADP-ribose) polymerase: an early marker of chemotherapy-induced apoptosis. *Cancer Res.* **53**, 3976–3985.

59. Lazebnik, Y. A., Kaufmann, S. H., Desnoyers, S., Poirier, G. G., and Earnshaw, W. C. (1994) Cleavage of poly(ADP-ribose) polymerase by a proteinase with properties like ICE. *Nature* **371**, 346–347.

60. Donato, N. J. and Perez M. (1998) Tumor necrosis factor-induced apoptosis stimulates p53 accumulation and p21WAF1 proteolysis in ME-180 cells. *J. Biol. Chem.* **273**, 5067–5072.

61. Levkau, B., Koyama, H., Raines, E. W., et al. (1998) Cleavage of p21Cip1/Waf1 and p27Kip1 mediates apoptosis in endothelial cells through activation of Cdk2: role of a caspase cascade. *Mol. Cell.* **1**, 553–563.

62. Waterhouse, N., Kumar, S., Song, Q., et al. (1996) Heteronuclear ribonucleoproteins C1 and C2, components of the spliceosome, are specific targets of interleukin 1beta-converting enzyme-like proteases in apoptosis. *J. Biol. Chem.* **271**, 29,335–29,341.

63. Tewari, M., Quan, L. T., O'Rourke, K., et al. (1995) Yama/CPP32 beta, a mammalian homolog of CED-3, is a CrmA-inhibitable protease that cleaves the death substrate poly(ADP-ribose) polymerase. *Cell* **81**, 801–809.

64. Widmann, C., Gibson, S., and Johnson, G. L. (1998) Caspase-dependent cleavage of signaling proteins during apoptosis. A turn-off mechanism for anti-apoptotic signals. *J. Biol. Chem.* **273**, 7141–7147.

65. Janicke, R. U., Walker, P. A., Lin, X. Y., and Porter, A. G. (1996) Specific cleavage of the retinoblastoma protein by an ICE-like protease in apoptosis. *EMBO J.* **15,** 6969–6978.

66. Zhou, B. B., Li, H., Yuan, J., and Kirschner, M. W. (1998) Caspase-dependent activation of cyclin-dependent kinases during Fas-induced apoptosis in Jurkat cells. *Proc. Natl. Acad. Sci. USA* **95,** 6785–6790.

67. Kayalar, C., Örd, T., Testa, M. P., Zhong, L. T., and Bredesen, D. E. (1996) Cleavage of actin by interleukin 1-converting enzyme to reverse DNase I inhibition. *Proc. Natl. Acad. Sci. USA* **93,** 2234–2238.

68. Mashima, T., Naito, M., Noguchi, K., Miller, D. K., Nicholson, D. W., and Tsuruo, T. (1997) Actin cleavage by CPP-32/apopain during the development of apoptosis. *Oncogene* **14,** 1007–1012.

69. Liu, X., Zou, H., Slaughter, C., and Wang, X. (1997) DFF, a heterodimeric protein that functions downstream of caspase-3 to trigger DNA fragmentation during apoptosis. *Cell* **89,** 175–184.

70. Sakahira, H., Enari, M., and Nagata, S. (1998) Cleavage of CAD inhibitor in CAD activation and DNA degradation during apoptosis. *Nature* **391,** 96–99.

71. Janicke, R. U., Ng, P., Sprengart, M. L., and Porter, A. G. (1998) Caspase-3 is required for alpha-fodrin cleavage but dispensable for cleavage of other death substrates in apoptosis. *J. Biol. Chem.* **273,** 15,540–15,545.

72. Kothakota, S., Azuma, T., Reinhard, C., et al. (1997) Caspase-3-generated fragment of gelsolin: effector of morphological change in apoptosis. *Science* **278,** 294–298.

73. Chan, W. H., Yu, J. S., and Yang, S. D. (2000) Apoptotic signalling cascade in photosensitized human epidermal carcinoma A431 cells: involvement of singlet oxygen, c-Jun N-terminal kinase, caspase-3 and p21-activated kinase 2. *Biochem. J.* **351,** 221–232.

74. Sebbagh, M., Renvoize, C., Hamelin, J., Riche, N., Bertoglio, J., and Breard, J. (2001) Caspase-3-mediated cleavage of ROCK I induces MLC phosphorylation and apoptotic membrane blebbing. *Nat. Cell Biol.* **3,** 346–352.

75. Coleman, M. L., Sahai, E. A., Yeo, M., Bosch, M., Dewar, A., and Olson, M. F. (2001) Membrane blebbing during apoptosis results from caspase-mediated activation of ROCK I. *Nat. Cell Biol.* **3,** 339–345.

76. van de Craen, M., de Jonghe, C., van den Brande, I., et al. (1999) Identification of caspases that cleave presenilin-1 and presenilin-2. Five presenilin-1 (PS1) mutations do not alter the sensitivity of PS1 to caspases. *FEBS Lett.* **445,** 149–154.

77. Chung, C. W., Song, Y. H., Kim, I. K., et al. (2001) Proapoptotic effects of tau cleavage product generated by caspase-3. *Neurobiol. Dis.* **8,** 162–172.

78. Goldberg, Y. P., Nicholson, D. W., Rasper, D. M., et al. (1996) Cleavage of huntingtin by apopain, a proapoptotic cysteine protease, is modulated by the polyglutamine tract. *Nat. Genet.* **13,** 442–449.

HSV Amplicon Vectors in Neuronal Apoptosis Studies

Kathleen A. Maguire-Zeiss, William J. Bowers, and Howard J. Federoff

1. Introduction

Specific cellular and temporal regulation of gene expression is a goal of many molecular studies. The study of programmed cell death requires cellular specificity, temporal regulation, as well as the interaction of a myriad of gene products. One way to regulate these interactions in an apoptotic cell is by specifically altering gene expression using a DNA transfer system. Several methods exist that are capable of delivering gene constructs into intact animals. DNA can be introduced into cells by direct DNA transfer using liposome-encapsulated DNA or viral vector systems which carry the gene of interest. An ex vivo approach can be implemented whereby cells are manipulated to produce the desired gene product and subsequently transferred to the animal. Transfer can also be accomplished using viral vector systems. In particular, transfer into the central nervous system and neurons is most commonly accomplished using various viral vector systems. Our laboratory and others have been developing herpes simplex virus (HSV) amplicon vectors, which are plasmid-based vectors that carry the gene of interest under the control of a specific promoter. In this chapter, we will review HSV amplicon vectors as a modality for gene transfer in two models of apoptosis, central nervous system (CNS) ischemia and cochlear degeneration. Improved helper-free amplicon methodology will be described, as well as advantages and disadvantages associated with this viral vector system.

From: *Neuromethods, Vol. 37: Apoptosis Techniques and Protocols, 2nd Ed.*
Edited by: A. C. LeBlanc @ Humana Press, Inc., Totowa, NJ

2. Herpes Simplex Virus Amplicon Vectors

Herpes Simplex Virus (HSV) amplicon vectors are plasmid-based vectors that take advantage of the natural neurotropism of the herpes virus. These vectors are capable of transducing a broad host of dividing and postmitotic cells. Efficient transduction of neuronal cells makes this vector system a powerful method to study neuron-specific phenomena. HSV amplicons possess a transgene capacity of approximately 130 kb, making them uniquely suited for the introduction of one or more foreign genes whose expression can be regulated by large and complex cellular promoters. This is in contrast to other vector systems such as retroviral-based vectors, recombinant adenoviral, and adeno-associated viral vectors, which have limited transgene capacity (4.5–8 kb). HSV amplicons are episomally expressed, which precludes random integration into the host genome, virtually eliminating potential oncogenic phenotypes. Lastly, HSV amplicons can now be packaged into infectious virus, which are devoid of contaminating cytotoxic helper virus (helper-free system) (Fig. 2). Using this method, we and others are able to produce relatively high-titer virus (10^8 T.U./mL) capable of efficient transgene expression in neurons and other cell types.

2.1. Structure of the HSV Amplicon Vector

The plasmid-based amplicon is essentially a eukaryotic expression plasmid that has been modified by the addition of an HSV origin of replication (ori) and cleavage/packaging sequence ("a" sequence; see Fig. 1). This highly versatile gene transfer system facilitates genetic manipulation owing to the fact that it has a large gene capacity (theoretical limit is 150 kb minus the size of the parental amplicon) (1–11). The amplicon vector concept arose from analysis of defective HSV particles that accumulated when HSV stocks were passaged at high multiplicities of infection (MOI) (12). These genomes were determined to contain an HSV origin of replication (ori) and pac "a" site. It was later shown that when the ori and pac were cloned into a plasmid, it could be replicated and packaged into virions (4). Conventional packaging required transfection of the amplicon into a "packaging" cell containing integrated complementing HSV IE gene(s) followed by superinfection with replication-defective helper virus. Analysis of these amplicon stocks revealed that they were composed of concatenated units of the original plasmid. Titers of these helper virus-containing

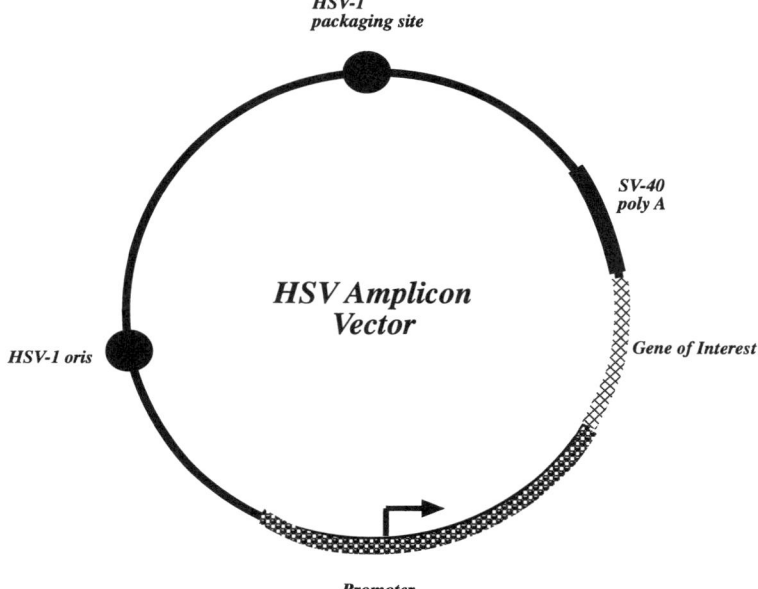

Fig. 1. Schematic of HSV amplicon vector. The prototypical vectors consist of the HSV-1 origin of replication and packaging site ("a"), a transcription unit composed of a cellular or viral promoter, the gene of interest, and an SV-40 polyadenylation site. The remaining pBR322 sequences support growth in *E. coli*.

amplicon stocks ranged from 10^6 to 10^8 and were capable of efficiently transferring genes into neurons in dissociated cell culture, organotypic slice culture, and in the intact brain (6,13–20).

2.2. Transgene Expression

When viral promoters are used to drive transgene expression, the result is transient high-level expression in a non-cell-specific manner. To begin to satisfy requirements for spatial and temporal control of gene expression, various iterations of the original amplicon design have been generated. Investigators have constructed amplicon vectors that contain multiple genes, regulated promoters, and cellular promoters (6,15–17). Inclusion of cellular promoters has been shown to confer long-term cell-specific expression. Specifically, Kaplitt et al. used the preproenkephalin promoter to demonstrate region-specific and long-term *lacZ* expression in the adult rat brain *(15)*. Our work using the 9.0-kb TH promoter fragment fused to an *Escherichia coli LacZ* reporter gene in a HSV amplicon vector (THlac) showed reporter gene expression preferentially in TH-positive substantia nigra dopaminergic neurons following infection *(17)*. HSV immediate-early (IE) 4/5 promoter-driven *LacZ* reporter (HSVlac) was used as a control in these experiments. Sprague-Dawley rats received unilateral stereotaxic injections of either THlac or HSVlac into the striatum or substantia nigra (SN). After 2 d or 10 wk, animals were sacrificed and processed for X-Gal histochemistry. *LacZ* expression driven by the TH promoter was prolonged (10 wk) and directed specifically to TH-positive neurons. *LacZ* expression was observed only when THlac virus was injected into the striatum, suggesting that retrograde transport facilitated efficient expression in TH-positive nigral neurons. Conversely, HSV promoter-driven amplicons were strongly but transiently expressed at the site of injection. Furthermore, *LacZ* expression was not detected in dopaminergic neurons. We and others have shown that amplicon genome persists when expression has been extinguished, suggesting a retention of amplicon that is transcriptionally inert *(16,21)*. The data outlined above suggest that cellular promoters can override some of the "silencing" and specificity issues related to the use of viral promoters.

Temporal control of gene expression can be addressed using a regulatable vector system. To accomplish this, inducible transcription units can be engineered into a vector expressing the gene of interest. For example, our laboratory developed a glucocorticoid-inducible HSV amplicon system that is capable of a 30-fold glucocorticoid-induction in transduced cell lines and a 50-fold induction in primary hepatocytes *(16)*. Briefly, a glucocorticoid-inducible transcription unit composed of reiterated steroid-responsive cis elements and the human growth hormone gene was inserted into an HSV amplicon vector. In these studies, there was a basal level of expression that was induced following steroid treatment. Current work by many groups is aimed at refining regulatable systems. The most common inducible systems belong to the following classes: tetracycline-regulated, ecdysone-inducible, antiprogestin-regulated, and dimerization-based regulation *(22)*. A recent advance has been made in the class of tetracycline-regulated systems. Strathdee et al. have developed an autoregulated bidirectional vector (pBIG) based on the original tetracycline-responsive system developed by Gossen and Bujard *(23,24)*. pBIG contains a bidirectional promoter that consists of the tetracycline-responsive herpes simplex virus thymidine kinase promoter (TK) and the tetracycline-responsive human cytomegalovirus immediate early promoter (CMV). In this system the weaker TK promoter is used to direct expression of the reversed tetracycline-responsive transactivator (rtTA) or the tetracycline-responsive transactivator (tTA) and the stronger CMV promoter directs the specific cDNA expression. Strathdee's group demonstrated that this system was able to regulate reporter gene expression tightly in both transient and stable transfectants but not in virus-transduced cells. Furthermore, this system can be switched between off and on states efficiently with the use of a tetracycline analog with no apparent cellular toxicity. Clearly, this type of regulated expression could be incorporated into a viral vector system with a large transgene capacity such as the HSV amplicon. Furthermore, gene transfer approaches capable of precise temporal regulation are an absolute necessity for future therapeutic studies.

3. Viral Vector-Based Gene Therapy in Apoptosis

Cell death by apoptosis is characterized by specific features such as cell shrinkage, membrane blebbing, chromatin condensation, DNA fragmentation, and phagocytosis without inflammation, which separate it from necrotic death *(25,26)*. Apoptosis is essential for normal development and maintenance of complex organisms. The cascade of apoptosis requires protein synthesis and is considered an active process. Although most apoptotic events follow a common death pathway, no one general antiapoptotic agent has been identified. A universal antiapoptotic agent may pose serious problems for an organism, since apoptotic death is necessary for normal development and a requirement in the prevention of certain diseases. With these caveats in mind, we will discuss some arenas where the regulation of apoptotic cellular events are being studied with gene transfer approaches. We will focus on neuroprotective strategies.

3.1. Stroke

Stroke occurs when cerebral blood flow is interrupted either by a thrombotic or embolic arterial occlusion or less often by a ruptured artery. When this interruption of blood flow reaches a critical point, ischemia results, causing cerebral tissue destruction. The degree of neurological damage is dependent in part on the location of the ischemic insult and the length of time the tissue is without oxygen and glucose. Following the initial ischemic event, neurological damage appears to be in two time domains, early and delayed. Early events result in a dense core of dead cells, which is due to necrotic cell death. The delayed cell death occurs hours to days later in the penumbra and is the result of apoptotic pathways (27). This apparent two-component process has been studied in animal models of stroke and has led to the identification of two therapeutic windows of opportunity to treat stroke patients. Practically speaking, interventions that will lessen or prevent the delayed apoptotic cell death process are likely to have the highest success rate in the treatment of stroke (28).

The cellular mechanisms underlying the cascade of events from cerebral ischemia to neurodegeneration are not completely understood but are similar to those seen in hypoxic and hypoglycemic states (29). In general, impeded blood flow results in a decrease of energy levels in the cell in the form of decreased ATP. Cell membranes increase their permeability, cells swell, cation pumps fail, and neuronal depolarization occurs, resulting in a rise in intracellular calcium. Calcium in turn regulates many cellular processes, including neurotransmitter release, activation of receptors, and second messenger systems (30,31). The cell's attempt to sequester calcium leads to mitochondrial damage and free-radical formation. Tissue pH drops as the brain continues to function in an anaerobic state, leading finally to cell death. Along the way, gene programs are turned on, some of which are neurotoxic and others neuroprotective. Oncogenes, such as those belonging to the bcl-2 gene family, are proposed to play an important role in the cell's ischemic response (32–35). Therapies designed to intervene using one or more of these genes would prove useful not only in the treatment of stroke but also for other neurological diseases that may share a common cell death pathway.

Bcl-2 is the founding member of a rapidly growing family of proteins involved in cell death regulation. First identified during the cloning of the t(14:18) chromosomal translocation breakpoint found in many follicular B-cell lymphomas, this 25- to 26-kDa protein is an integral membrane component that associates with mitochondrial, endoplasmic reticulum, and nuclear membranes (36–43). Changes in bcl-2 expression have been demonstrated in various cell death paradigms (33,44–52). The unique function of this oncogene is that Bcl-2 appears to prevent cell death by promoting cell survival (antiapoptotic), whereas other genes, including some *bcl-2*–related family members, are known to augment cell death [pro-apoptotic *(40,53,54)*]. For example, p53 overexpression transactivates Bax, which leads to caspase-3 activation and apoptotic cell death *(55–58)*. Bcl-2 family members can protect cells from Bax-induced apoptosis.

Using viral vector-based gene transfer, several laboratories have shown the effectiveness of neuroprotective genes in rescuing cells from apoptosis. Martinou and colleagues demonstrated that overexpression of Bcl-2 in transgenic mice under conditions of experimental ischemia resulted in approximately 50% reduction in infarct area *(51)*. In these studies, the middle cerebral artery was occluded for 7 d, at which time the infarct area was determined. Both *bcl-2* transgenics and wild-type littermates demonstrated a focal ischemic core of necrosis, but the diffuse infarct area was significantly smaller in the transgenic animals, suggesting a cell survival effect of Bcl-2. Our laboratory has provided further evidence for Bcl-2's neuroprotective role in ischemia *(50)*. Using a defective herpes simplex virus-1 vector (HSV-1) expressing *bcl-2*, a protective effect of *bcl-2* expression in rat cortex was demonstrated when the gene was delivered prior to the induction of focal ischemia. In these studies, HSVbcl2 or HSVlac (*Escherichia coli lacZ*) were injected into two sites in rat cerebral cortex 24 h before induction of focal ischemia. Ischemia was achieved by tandem permanent occlusion of the right middle cerebral artery and ipsilateral common carotid artery. Expression of *bcl-2* was confirmed by immunohistochemistry. Twenty-four hours after occlusion, local ischemic damage was determined by staining with 2% 2,3,5-triphenyltetrazolium chloride. Viable tissue was significantly increased in HSVbcl2-treated rats with protection localized to the injection sites. In a model of transient global ischemia, Mongolian gerbil CA1 neurons were protected from delayed neuronal death by the prior administration of HSVbcl2 *(59)*. Adenovirus-mediated transfer of another Bcl family member, Bcl-X_L, also protects neurons from Bax induced apoptosis *(57)*. Utilizing PC12 cells, these authors demonstrated that Bcl-X_L was more effective in preventing the Bax-induced cell death than Bcl-2, which may reflect the normal higher level expression of Bcl-X_L in mature neurons. Another neuroprotective gene, glial cell line-derived neurotrophic factor (GDNF), is also important in rescuing neurons from cell death. Recombinant adeno-associated virus vectors expressing GDNF have been shown to effectively

decrease infarct volume and maintain neuron number following transient bilateral common carotid artery and unilateral middle cerebral
artery ligation *(60)*. Taken together, these studies point to a protective role for *gdnf* and *bcl-2* family members in stroke and suggest
the feasibility of gene therapy for this disease.

Since stroke paradigms share many of the same features as hypoxic
and hypoglycemic states, we can postulate other areas where gene
therapy will be a valuable tool in preventing apoptotic cell death.
Under hypoxic and hypoglycemic states, cells initiate a variety of
adaptive events. Two gene products found to be upregulated during these events, hypoxia-inducible factor-1α (HIF-1α) and ARNT,
are members of the Per-ARNT-Sim (PAS) family of transcription
factors and will heterodimerize and translocate to the nucleus
following hypoxia *(61,62)*. The HIF-1α complex then regulates the
expression of adaptive genes such as erythropoietin by binding
to specific hypoxia-responsive enhancer elements *(63,64)*. Prolonged
hypoxia has an apparent contrary effect in that HIF-1α is stabi-
lized and activates cell death in a p53-dependent manner *(20,65)*.
Studies from our laboratory using HSV amplicon gene transfer of
a dominant-negative form of HIF-1α (HIFdn) into cortical neuronal cultures showed reduced delayed neuronal death following
hypoxic stress *(20)*. Therefore, p53 and HIF-1α appear to conspire
to promote apoptotic cell death following ischemia and provide
us with yet another area for gene therapeutic intervention.

3.2. Cochlear Degeneration

The ear represents a particularly amenable target for viral vector-mediated gene therapy *(66)*. Since most humans do not have
a patent cochlear aqueduct, the ear is a closed system with little
chance of virus spread to other organs. Mechanical delivery to
the inner ear can be achieved via a small cochleostomy. Virus is
then able to access inner ear tissues such as the sprial ganglion
and organ of Corti perilymphatic spaces. Spiral ganglion neurons
(SGN) connect hair cells of the organ of Corti to the central nervous system via the cochleovestibular nerve (VIIIth). SGN require
a central target (cochlear nucleus) and peripheral target (organ of
Corti) for survival *(67–73)*.

First postulated by Ramon y Cajal and later confirmed by Lefebvre
et al., a diffusable target-derived survival factor exists for these
neurons *(74)*. In early postnatal cochlea, brain-derived neurotrophic factor (BDNF) and neurotrophin-3 (NT-3) are expressed in

both inner and outer hair cells. More specifically, BDNF has been localized to sensory cells of both auditory and vestibular organs, while NT-3 is expressed in inner hair cells and to a lesser extent in outer hair cells (75). Receptors for these neurotrophins have been localized to SGNs (75,76). SGNs are classified as type I or type II based on their specific functions. Type I SGNs deliver sound signal information from the inner hair cells to the brain, while type II SGNs innervate outer hair cells (77). Both type I and type II SGNs require neurotrophin-3 (NT-3) for survival. Severing the cochleovestibular nerve, aminoglycoside ototoxicity of the hair cells and cisplatin administration all lead to a secondary degeneration of the spiral ganglion neurons (6,68,70,78). In turn, destruction of hair cells and SGN degeneration result in hearing loss. Some forms of deafness are alleviated by cochlear implants, which are surgically implanted hearing aids that are placed directly in the scala tympani of the cochlea. These implants send electrical impulses to the cochlear nerve but require a population of intact SGN (69,79,80).

In an attempt to maintain functional neurons, neurotrophins have been delivered to auditory ganglia via vector-mediated delivery methods. In one study from our laboratory, HSV amplicon vectors expressing BDNF were used to transduce SGN cultures (HSVbdn flac) (6). The expression of BDNF elicited robust neuritic process outgrowth, which was comparable to the addition of exogenous BDNF. This initial gene therapy study demonstrated the effectiveness of the HSV amplicon vector system to deliver genes to the SGNs. Further work has demonstrated that HSV amplicon-delivered BDNF prevents SGN degeneration in a mouse model of aminoglycoside ototoxicity (81). Ototoxicity is also a serious concern with some chemotherapeutic treatments such as cisplatin and is in fact a major dose-limiting side effect for this cancer therapy (82–85). Localized delivery of exogenous NT-3 has proven effective in preventing SGN loss following cisplatin treatment, but in vivo treatment requires an efficient delivery system (86,87). Toward that end, our laboratory has developed an HSV amplicon vector that expresses a c-Myc-tagged NT-3 chimera (HSVnt-3myc) (78). Transduction of cultured murine cochlear explants with HSVnt-3myc resulted in expression of both NT-3 mRNA and protein. NT-3 transduction increased the density of SGN fibers that projected toward hair cells as measured by immunocytochemistry for neurofilament-200. To examine the ability of NT-3 to prevent SGN loss during cisplatin treatment, explants were transduced with HSVnt-3myc and 48 h later exposed to varying concentrations of cisplatin. Immunocyto-chemical analysis demonstrated that NT-3 overexpression increased the number of neurites and rescued SGNs. These data suggest that an HSV amplicon vector expressing a neurotrophin is a promising methodology to effect biologically meaningful gene expression in vivo. As mentioned previously, expression can be regulated by the choice of promoter and in vivo delivery could be accomplished using a miniosmotic pump and microcatheter inserted through the lateral wall of the cochlea or round window (88,89).

In summary, stroke and cochlear degeneration represent excellent models for HSV amplicon-mediated gene expression therapy to abrogate cell death. This vector system is a versatile platform, as it allows for the incorporation of large transgenes, specific promoters, and neurotropic expression.

4. Methods

As vector improvement is an ongoing goal in our laboratory as well as others, the methods for HSV amplicon virus production are constantly changing. Here, we will describe the most current production protocol (90,91).

4.1. Cell Culture

Baby hamster kidney (BHK) cells are maintained in Dulbecco's modified Eagle's medium high glucose (DMEM) supplemented with 10% fetal bovine serum (FBS), 100-U/mL penicillin, and 100-μg/mL streptomycin (Life Technologies). The NIH-3T3 mouse fibroblast cell line, originally obtained from American Type Culture Collection, is maintained in DMEM supplemented with FBS, 100-U/mL penicillin, and 100-μg/mL streptomycin.

4.2. Amplicon Vector DNA

The gene of interest is subcloned into an HSV amplicon vector (see Fig. 1) containing an HSV-1 origin of DNA replication (oriS) that supports the replication of vector DNA and the HSV-1 "a" sequence that contains the packaging site responsible for subsequently packaging vector DNA into HSV-1 particles (13). The promoter used to drive gene expression is subcloned 5-prime of the gene of interest. Amplicon DNA is purified using standard DNA purification procedures.

4.3. BAC DNA Preparation

pBAC-V2 DNA represents the entire 152-kb genome of HSV-1 without the *pac* signals cloned into a bacterial artificial chromosome *(92,93)*. When co-transfected with amplicon DNA into BHK cells, the amplicon DNA is efficiently packaged. This system eliminates the replication-competent helper virus and associated immunogenicity. A more recent iteration of this BAC system was recently reported by Saeki et al., which includes a deletion of the ICP27 gene and the addition of the ICP0 "stuffer" sequence in an attempt to eliminate the helper component and increase titers *(94)*.

Initially, 4 L of LB containing 12.5-µg/µL chloramphenicol is inoculated with pBAC-V2 glycerol stock and grown overnight in a 37°C shaking incubator. pBAC-V2 DNA is purified using the following modified method based on the Endo-Free Plasmid Mega Kit protocol from Qiagen (cat. no. 12381, and Endo-Free Plasmid Buffer Set, cat. no. 19048). Bacteria is pelleted by centrifugation for 20 min at 2700g and resuspended in 100 mL of P1 buffer. Following the addition of 100 mL of P2 buffer, the mixture is allowed to incubate for 5 min at room temperature. Prechilled P3 buffer (100 mL) is added, and incubation continues for an additional 30 min on ice. At the end of the incubation, the mixture is centrifuged at 20000g for 30 min at 4°C. The supernatant is passed through two layers of prewetted cheesecloth. The supernatant is further cleared by passing it through prewetted filter paper to prevent column clogging (S&S grade 588, cat. no. 10319344, or grade 606, cat. no. 03890; Whatman grade 59728435). ER buffer (30 mL) is added to the cleared supernatant and the mixture is inverted 10 times and allowed to incubate on ice for 30 min. During this incubation, two Qiagen-tip 2500 columns are equilibrated by applying 35 mL QBT buffer to each. Once equilibrated, half of the cleared lysate is applied to each column and allowed to enter the resin by gravity flow. The columns are washed with 220 mL of QC buffer and DNA eluted with 35 mL (5 × 7 mL) of prewarmed (56°C) QN buffer. The eluted DNA is precipitated with 0.7 vol of room-temperature isopropanol and centrifuged immediately at >15,000g for 30 min at 4°C. The DNA pellets are rinsed with 2 mL of endotoxin-free 70% ethanol, centrifuged for 10 min at 15,000g, and resuspended in a total of 2 mL endotoxin-free TE (10 mM Tris-HCl, pH 7.5, 1 mM EDTA). Pellets are allowed to resuspend at 4°C overnight, followed by gentle rocking for 2 h at room temperature. DNA concentrations

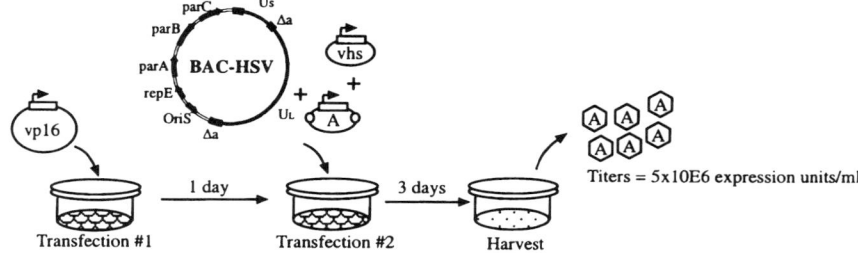

Fig. 2. Schematic of BAC-based amplicon packaging. A vp16-express-ing plasmid is transfected into the packaging cell line followed 24 h later by contransfection with BAC-HSV, a vhs-expressing plasmid and HSV amplicon vector. After 3 d, amplicon virus is harvested and purified as outlined under Subheading 4.

are determined by absorbance readings at a wavelength of 260 nm. DNA is stored at −20°C.

4.4. Helper Virus-Free HSV Amplicon Packaging

On the day prior to transfection, 2×10^7 BHK cells are seeded in a T-150 flask and incubated overnight at 37°C (Fig. 2). The day of transfection, 1.8 mL Opti-MEM (Gibco-BRL, Bethesda, MD), 25 µg of pBAC-V2 DNA, and 7 µg amplicon vector DNA are combined in a sterile polypropylene tube. Seventy microliters of Lipofect-amine Plus Reagent (Gibco-BRL) is added over a period of 30 s to the DNA mix and allowed to incubate at 22°C for 20 min. In a separate tube, 100 µL Lipofectamine (Gibco-BRL) is mixed with 1.8 mL Opti-MEM and also incubated at 22°C for 20 min. Follow-ing the incubations, the contents of the two tubes are combined over a period of 30 s, then incubated for an additional 20 min at 22°C. During this second incubation, the media in the seeded T-150 flask is removed and replaced with 14 mL Opti-MEM. The transfection mix is added to the flask and allowed to incubate at 37°C for 5 h. The transfection mix is then diluted with an equal volume of DMEM plus 20% FBS, 2% penicillin/streptomycin, and 2 mM hexamethylene bis-acetamide (HMBA), and incubated over-night at 34°C. The following day, the medium is removed and replaced with DMEM plus 10% FBS, 1% penicillin/streptomycin, and 2 mM HMBA. The packaging flask is incubated an additional

3 d before virus is harvested and stored at –80°C until purification. Viral preparations are subsequently thawed, sonicated, clarified by centrifugation at 1000*g* for 15 min, and concentrated by ultracentrifugation through a 30% sucrose cushion at 107,000*g* for 45 min (detailed below). Viral pellets are resuspended in 100 µL PBS and stored at –80°C until use. Vectors are titered as described previously, using either transduction-based or expression-based titering methods *(91)*.

4.5. BAC Virus Concentration

Thaw previously harvested virus quickly at 37°C and transfer tubes to ice. Sonnicate on ice 3 times at setting 8, alternating 30 s on/30 s off, using a Missonix cup sonnicator. Centrifuge at 1300*g* for 10 min. Remove viral supernatant to a fresh 50-mL conical and centrifuge again at 3000*g* for 10 min. Meanwhile, rinse 37-mL Sorvall tubes with 70% ethanol 2 times to remove residue. Air-dry by inverting tubes, remove any remaining alcohol with a sterile Q-tip. Once dry, add 10–15 mL of DMEM to the tubes and underlay with 7 mL cold, sterile 30% sucrose in PBS. Carefully overlay the viral supernatant and top off each tube with cold DMEM to within 0.25 cm of the tube rim. Place tubes in prechilled ultracentrifuge swinging buckets using forceps sterilized with 70% ethanol. Centrifuge in a prechilled ultracentrifuge for 2 h at 107,000*g* at 4°C. Decant the supernatant, remove residual medium with a sterile Q-tip, and add PBS containing Ca^{2+}, Mg^{2+} to the viral pellet (use 100 µL/T-150). Incubate on ice for 30 min to soften pellet and resuspend by gently pipetting, to break up viral clumps. Aliquot and freeze on Dry Ice. Store at –80°C until use.

4.6. Viral Titering

Amplicon titers are determined either by assessing the number of cells expressing a given transduced gene using X-gal histochemistry or by assessing the number of transduced viral genomes using a Perkin Elmer 7700 quantitative polymerase chain reaction (PCR)-based method. For expression titers, 5 µL of concentrated amplicon stock is incubated with confluent monolayers of NIH 3T3 cells. Following a 24-h incubation, cells are fixed with 1% glutaraldehyde for 5 min at room temperature and processed for X-gal histochemistry to detect the *lacZ* transgene product. Positively stained blue cells are counted, and expression titer is calculated and represented as blue-forming units per milliliter (bfu/mL). For transduction titers, infection of confluent monolayers is carried out as described above, but following the overnight incubation, monolayers are processed for total DNA. Briefly, cells are lysed in 100 mM potassium phosphate pH 7.8 and 0.2% Triton X-100. An equal volume of 2X digestion buffer (0.2 M NaCl, 20 mM Tris-Cl, pH 8, 50 mM EDTA, 0.5% sodium dodecyl sulfate (SDS), 0.2-mg/mL proteinase K) is added to the remainder of the lysate and the sample is incubated at 56°C for 4 h. Samples are processed further by one phenol/chloroform, one chloroform extraction, and a final ethanol precipitation. Total DNA is quantitated using a spectrophotometer, and 50 ng of DNA is analyzed in a PE7700 quantitative PCR reaction using a designed *lacZ*-specific primer/probe combination multiplexed with an 18S rRNA-specific primer/probe set. The *lacZ* probe sequence is

                5'-ACCCCGTACGTCTTCCCGAGCG-3';

the *lacZ* sense primer sequence is

                5'- GGGATCTGCCATTGTCAGACAT-3';

and the *lacZ* antisense primer sequence is

                5'- TGGTGTGGGCCATAATTCAA-3';

The 18S rRNA probe sequence is

                5'-TGCTGGCACCAGACTTGCCCTC-3';

the 18S sense primer sequence is

                5'-CGGCTACCACATCCAAGGAA-3';

and the 18S antisense primer sequence is

                5'-GCTGGAATTACCGCGGCT-3'.

4.7. TaqMan Quantitative PCR System

Each 25-µL PCR sample contains 2.5 µL of purified DNA, 900 nM of each primer, 50 nM of each probe, and 12.5 µL of 2X Perkin-Elmer Master Mix. Following a 2-min 50°C incubation and 2-min 95°C denaturation step, the samples are subjected to 40 cycles of 95°C for 15 s and 60°C for 1 min. Fluorescent intensity of each sample is detected automatically during the cycles by the Perkin-Elmer Applied Biosystem Sequence Detector 7700 machine. Each PCR run includes the following: no-template control samples, positive control samples consisting of either amplicon DNA (for *lacZ*) or cellular genomic DNA (for 18S rRNA), and two standard-curve dilution series (for *lacZ* and 18S). Following the PCR run, "real-time" data are analyzed using Perkin-Elmer Sequence Detector Software, Version 1.6.3, and the aforementioned standard curves. Precise quantities of starting template are determined for each titering sample and results are expressed as numbers of vector genomes per milliliter of original viral stock.

4.8. Transduction of Cells

Stable cell lines, primary cell cultures, and tissue explants are effectively transduced with these HSV amplicon vectors *(78,90)*. The amount of amplicon virus used needs to be determined for each cell type. In general, cell culture medium volume is reduced and amplicon virus added to the cultures. The cultures are gently swirled every 15 min for 1–2 h. Medium is removed and replaced with fresh medium. For primary cells or cells that require conditioned medium, the original medium prior to transduction is retained and used for re-feeding.

HSV amplicon vectors can also effectively transduce cells in vivo *(90)*. Our laboratory has extensive expertise in the stereotactic delivery of HSV amplicon virus to the rodent brain *(90,95–97)*. In general, mice are anesthetized with avertin at a dose of 0.6 mL per 25 g body weight. After positioning in an ASI murine stereotactic apparatus, the skull is exposed via a midline incision and burr holes drilled over the appropriate coordinates. A 33-gage steel needle is lowered to the appropriate depth over a 5-min time period. Amplicon virus is than infused using a microprocessor-controlled pump over a 10-min time period (UltraMicroPump, World Precision Instruments, Sarasota Springs, FL; detailed in ref. 97). The injector unit is mounted on a precision small animal stereotactic frame micromanipulator at a 90° angle using a mount for the injector (ASI Instruments, Warren, MI). The microprocessor-controlled pump is completely programmable, allowing for a precise delivery rate. The volume of virus injected is dependent on the site of injection, but in our experience for mouse brain the volume is limited to between 1 and 1.5 µL. The needle is then manually removed over a 10-min time period and the skin incision sutured closed. Animals generally recover within 45 min of surgery.

5. Advantages and Disadvantages of the HSV Amplicon System

As with all current gene therapy systems, there are advantages and disadvantages to using the HSV virus-free amplicon system (for review, *see* ref. *93*). The advantages of this system lie in: (1) the amplicon's inherent large transgene capacity; (2) the amplicon's ability to infect both dividing and nondividing cells, specifically neurons; and (3) the helper-free amplicon's low immunogenicity. This system has been effective in achieving transgene expression for a relatively long time (months to 1 yr), but this is the area that will require the most improvement. Stable and long-term expression is an ongoing goal of HSV amplicon therapy and is beginning to be addressed with the use of specific cellular promoters, specific insulting DNA elements, and nuclear matrix attachment sequences. Due to the large transgene capacity of the HSV amplicon, such regulatory elements can be easily incorporated into this vector. Overall, the HSV helper-free amplicon vector represents an important component in our armament of pre-clinical gene therapy strategies for the study of many important biological phenomena including apoptosis.

Acknowledgments

This work was supported by an AFAR Research Grant to WJB and NIH R01-NS36420A to HJF.

References

1. Federoff, H. J., Geschwind, M. D., Geller, A. I., and Kessler, J. A. (1992) Expression of nerve growth factor in vivo from a defective herpes simplex virus 1 vector prevents effects of axotomy on sympathetic ganglia. *Proc. Natl. Acad. Sci. USA* **89**(5), 1636–1640.
2. Frenkel, N. (1981) Defective interfering herpesviruses, in *The Human Herpesviruses—An Interdisciplinary Prospective* (Nahmias, A., Dowdle, W., and Scchinazy, R., eds.), Elsevier-North Holland, New York, pp. 91–120.
3. Frenkel, N., Spaete, R. R., Vlazny, D. A., Deiss, L. P., and Locker, H. (1982) The herpes simplex virus amplicon—a novel animal-virus cloning vector, in *Eucaryotic Viral Vectors* (Gluzman, Y., ed.), Cold Spring Harbor Laboratory, Cold Spring Harbor, NY, pp. 205–209.
4. Geller, A. I. and Breakefield, X. O. (1988) A defective HSV-1 vector expresses *Escherichia coli* β-galactosidase in cultured peripheral neurons. *Science* **241,** 1667–1669.

5. Geller, A. I., Keyomarsi, K., Bryan, J., and Pardee, A. B. (1990) An efficient deletion mutant packaging system for defective herpes simplex virus vectors: potential applications to human gene therapy and neuronal physiology. *Proc. Natl. Acad. Sci. USA* **87,** 8950–8954.

6. Geschwind, M. D., Hartnick, C. J., Liu, W., Amat, J., Van De Water, T. R., and Federoff, H. J. (1996) Defective HSV-1 vector expressing BDNF in auditory ganglia elicits neurite outgrowth: model for treatment of neuron loss following cochlear degeneration. *Hum. Gene Ther.* **7,** 173–182.

7. Geschwind, M. D., Lu, B., and Federoff, H. J. (1994) Expression of neurotrophic genes from herpes simplex virus type 1 vectors: modifying neuronal phenotype. *Meth. Neurosci.* **21,** 462–482.

8. Lawrence, M., Ho, D., Dash, R., and Sapolsky, R. (1995) Herpes simplex virus overexpressing the glucose transporter gene protect against seisureinduced neuron loss. *Proc. Natl. Acad. Sci. USA* **92,** 7247–7251.

9. Spaete, R. R. and Frenkel, N. (1982) The herpes simplex virus amplicon: a new eucaryotic defective-virus cloning-amplifying vector. *Cell* **30,** 305–310.

10. Spaete, R. R. and Frenkel, N. (1985) The herpes simplex virus amplicon: analyses of *cis*-acting replication functions. *Proc. Natl. Acad. Sci. USA* **82,** 694–698.

11. Stow, N. D. and McMonagle, E. (1982) Propagation of foreign DNA sequences linked to a herpes simplex virus origin of replication, in *Eucaryotic Viral Vectors* (Gluzman, Y., ed.), Cold Spring Harbor Laboratory, Cold Spring Harbor, NY, pp. 199–204.

12. Frenkel, N., Locker, H., Batterson, W. Hayward, G. S., and Roizman, B. (1976) Anatomy of herpes simplex virus DNA. VI. Defective DNA originates from the S component. *J. Virol.* **20**(2), 527–531.

13. Geller, A. I. and Federoff, H. J. (1991) The use of HSV-1 vectors to introduce heterologous genes into neurons: implication for gene therapy, in *Human Gene Transfer* (Cohen-Haguenauer, O. and Boiron, M., eds.), Colloque INSERM/John Libbey Eurotext, Montrouge, France, pp. 63–73.

14. Geschwind, M. D., Kessler, J. A., Geller, A. I., and Federoff, H. J. (1994) Transfer of the nerve growth factor gene into cell lines and cultured neurons using a defective herpes simplex virus vector. Transfer of the NGF gene into cells by a HSV-1 vector. *Mol. Brain Res.* **24,** 327–335.

15. Kaplitt, M. G., Leone, P., Samulski, R. J., et al. (1994) Long-term gene expression and phenotypic correction using adeno-associated virus vectors in the mammalian brain. *Nat. Genet.* **8**(Oct.), 148–154.

16. Lu, B. and Federoff, H. J. (1995) Herpes simplex virus type 1 amplicon vectors with glucocorticoid-inducible gene expression. *Hum. Gene Ther.* **6,** 421–430.

17. Jin, B. K., Belloni, M., Conti, B., et al. (1996) Prolonged *in vivo* gene expression driven by a tyrosine hydroxylase promoter in a defective herpes simplex virus amplicon vector. *Hum. Gene Ther.* **7,** 2015–2024.

18. Liu, X., Kim, C. N., Yang, J., Jemmerson, R., and Wang, X. (1996) Induction of apoptotic program in cell-free extracts: requirement for dATP and cytochrome c. *Cell* **86,** 147–157.

19. Brooks, A. I., Muhkerjee, B., Panahian, N., Cory-Slechta, D., and Federoff, H. J. (1997) Nerve growth factor somatic mosaicism produced by herpes virus-directed expression of *cre* recombinase. *Nat. Biotechnol.* **15,** 57–62.

20. Halterman, M. W., Miller, C. C., and Federoff, H. J. (1999) Hypoxia-inducible factor-1α mediates hypoxia-induced delayed neuronal death that involves p53. *J. Neurosci.* **19**(16), 6818–6824.

21. During, M. J., Naegele, J. R., O'Malley, K. L., and Geller, A. I. (1994) Long-term behavioral recovery in Parkinsonian rats by an HSV vector expressing tyrosine hydroxylase. *Science* **266,** 1399–1403.

22. Harvey, D. M. and Caskey, C. T. (1998) Inducible control of gene expression: prospects for gene therapy. *Curr. Opin. Chem. Biol.* **2**(4), 512–518.

23. Gossen, M. and Bujard, H. (1992) Tight control of gene expression in mammalian cells by tetracycline-responsive promoters. *Proc. Natl. Acad. Sci. USA* **89**(12), 5547–5551.

24. Strathdee, C. A., McLeod. M. R., and Hall. J. R. (1999) Efficient control of tetracycline-responsive gene expression from an autoregulated bi-directional expression vector. *Gene* **229**(1–2), 21–29.

25. Kerr, J. F., Wyllie. A. H., and Currie, A. R. (1972) Apoptosis: a basic biological phenomenon with wide-ranging implications in tissue kinetics. *Br. J. Cancer* **26**(4), 239–257.

26. Michel, P. P., Lambeng, N., and Ruberg, M. (1999) Neuropharmacologic aspects of apoptosis: significance for neurodegenerative diseases. *Clin. Neuropharmacol.* **22**(3), 137–150.

27. Emerich, D. and Bartus, R. (1999) Intracellular events associated with cerebral ischemia, in *Stroke Therapy: Basic, Preclinical, and Clinical Directions* (Miller, L., ed.), Wiley-Liss, New York, pp. 195–218.

28. Sapolsky, R. M. and Steinberg, G. K. (1999) Gene therapy using viral vectors for acute neurologic insults. *Neurology* **53**(9), 1922–1931.

29. Hatterman, M. W. and Federoff, H. J. (1999) Hif-1a and p53 promote hypoxia-induced delayed neuronal death in models of CNS ischemia. *Exp. Neurol.* **159,** 65–72.

30. Siesjo, B. K. (1981) Cell damage in the brain. *J. Cereb. Blood Flow Metab.* **1,** 155–185.

31. Koroshetz, W. J. and Moskowitz, M. A. (1996) Emerging treatments for stroke in humans. *TiPS* **17,** 227–233.

32. Collaco-Moraes, Y., Aspey, B. S., de Belleroche, J. S., and Harrison, M. J. G. (1994) Focal ischemia causes an extensive induction of immediate early genes that are sensitive to MK-801. *Stroke* **25,** 1855–1861.

33. Gillardon, F., Lenz, C., Waschke, K. F., et al. (1996) Altered expression of Bcl-2, Bcl-X, Bax, and c-Fos colocalizes with DNA fragmentation and ischemic cell damage following middle cerebral artery occlusion in rats. *Mol. Brain Res.* **40,** 254–260.

34. Kogure, K. and Kato, H. (1993) Altered gene expression in cerebral ischemia. *Stroke* **24**(12), 2121–2127.

35. Xiang, H., Kinoshita, Y., Knudson, C. M., Korsmeyer, S. J., Schwartzkroin, P. A., and Morrison, R. S. (1998) Bax involvement in p53-mediated neuronal cell death. *J. Neurosci.* **18**(4), 1363–1373.

36. Bakhshi, A., Jensen, J. P., Goldman, P., et al. (1985) Cloning the chromosomal breakpoint of t(14;18) human lymphomas: clustering around J$_H$ on chromosome 14 and near a transcriptional unit on 18. *Cell* **41**(3), 899–906.

37. Cleary, M. L. and Sklar, J. (1985) Nucleotide sequence of a t(14;18) chromosomal breakpoint in follicular lymphoma and demonstration of a break-

point-cluster region near a transcriptionally active locus on chromosome 18. *Proc. Natl. Acad. Sci. USA* **82**(21), 7439–7443.

38. Hockenbery, D., Nunez, G., Milliman, C., Schreiber, R. D., and Korsmeyer, S. J. (1990) Bcl-2 is an inner mitochondrial membrane protein blocks programmed cell death. *Nature* **348**, 334–336.

39. Kajewski, S., Tanaka, S., Takayama, S., Schibler, M. J., Fenton, W., and Reed, J. C. (1993) Investigation of the subcellular distribution of the *bcl-2* oncoprotein: residence in the nuclear envelope, endoplasmic reticulum, and outer mitochondrial membranes. *Cancer Res.* **53**(19), 4701–4714.

40. Merry, D. E. and Korsmeyer, S. J. (1997) Bcl-2 gene family in the nervous system. *Annu. Rev. Neurosci.* **20**, 245–267.

41. Monaghan, P., Robertson, D., Amos, T. A. S., Dyer, M. J. S., Mason, D. Y., and Greaves, M. F. (1992) Ultrastructural localization of bcl-2 protein. *J. Histochem. Cytochem.* **40**(12), 1819–1825.

42. Tsujimoto, Y., Gorham, J., Cossman, J., Jaffe, E., and Croce, C. M. (1985) The t(14;18) chromosome translocations involved in B-cell neoplasms result from mistakes in VDJ joining. *Science* **229**, 1390–1393.

43. Reed, J. C. (1997) Double identity for proteins of the Bcl-2 family. *Nature* **387**, 773–776.

44. Allsopp, T. E., Wyatt, S., Paterson, H. F., and Davies, A. M. (1993) The proto-oncogene bcl-2 can selectively rescue neurotrophic factor-dependent neurons from apoptosis. *Cell* **73**, 295–307.

45. Asahi, M., Hoshimaru, M., Uemura, Y., et al. (1997) Expression of interleukin-1β converting enzyme gene family and *bcl-2* gene family in the rat brain following permanent occlusion of the middle cerebral artery. *J. Cereb. Blood Flow Metabol.* **17**, 11–18.

46. Behl, C., Hovey, L. III, Krajewski, S., Schubert, D., and Reed, J. C. (1993) Bcl-2 prevents killing of neuronal cells by glutamate but not by amyloid beta protein. *Biochem. Biophys. Res. Commun.* **197**(2), 949–956.

47. Chen, J., Graham, S. H., Nakayama, M., et al. (1997) Apoptosis repressor genes Bcl-2 and Bcl-x-long are expressed in the rat brain following global ischemia. *J. Cereb. Blood Flow Metabol.* **17**, 2–10.

48. Honkaniemi, J., Massa, S. M., Breckinridge, M., and Sharp, F. R. (1996) Global ischemia induces apoptosis-associated genes in hippocampus. *Mol. Brain Res.* **42**, 79–88.

49. Kane, D. J., Sarafian, T. A., Anton, R., et al. (1993) Bcl-2 inhibition of neural death: decreased generation of reactive oxygen species. *Science* **262**, 1274–1277.

50. Linnik, M. D., Zahos, P., Geschwind, M. S., and Federoff, H. J. (1995) Expression of *bcl-2* from a defective herpes simplex virus-1 vector limits neuronal death in focal cerebral ischemia. *Stroke* **26**, 1670–1675.

51. Martinou, J.-C., Dubois-Dauphin, M., Staple, J. K., et al. (1994) Overexpression of BCL-2 in transgenic mice protects neurons from naturally occurring cell death and experimental ischemia. *Neuron* **13**, 1017–1030.

52. Zhong, L.-T., Sarafian, T., Kane, D. J., et al. (1993) Bcl-2 inhibits death of central neural cells induced by multiple agents. *Proc. Natl. Acad. Sci. USA* **90**, 4533–4537.

53. McDonnell, T. J., Deane, N., Platt, F. M., et al. (1989) *Bcl-2*-immunoglobulin transgenic mice demonstrate extended B cell survival and follicular lymphoproliferation. *Cell* **57**, 79–88.

54. Vaux, D. L., Weissman, I. L., and Kim, S. K. (1992) Prevention of programmed cell death in *Caenorhabditis elegans* by human *bcl-2*. *Science* **258**, 1955–1957.

55. Eizenberg, O., Faber-Elman, A., Gottlieb, E., Oren, M., Rotter, V., and Schwartz, M. (1996) p53 plays a regulatory role in differentiation and apoptosis of central nervous system-associated cells. *Mol. Cell Biol.* **16**(9), 5178–5185.

56. Jordan, J., Galindo, M. F., Prehn, J. H., et al. (1997) p53 expression induces apoptosis in hippocampal pyramidal neuron cultures. *J. Neurosci.* **17**(4), 1397–1405.

57. Shinoura, N., Satou, R., Yoshida, Y., Asai, A., Kirino, T., and Hamada, H. (2000) Adenovirus-mediated transfer of Bcl-X(L) protects neuronal cells from Bax-induced apoptosis. *Exp. Cell Res.* **254**(2), 221–231.

58. Slack, R. S., Belliveau, D. J., Rosenberg, M., et al. (1996) Adenovirus-mediated gene transfer of the tumor suppressor, p53, induces apoptosis in postmitotic neurons. *J. Cell Biol.* **135**(4), 1085–1096.

59. Antonawich, F. J., Federoff, H. J., and Davis, J. N. (1999) BCL-2 transduction, using a herpes simplex virus amplicon, protects hippocampal neurons from transient global ischemia. *Exp. Neurol.* **156**(1), 130–137.

60. Tsai, T. H., Chen, S. L., Chiang, Y. H., et al. (2000) Recombinant adeno-associated virus vector expressing glial cell line-derived neurotrophic factor reduces ischemia-induced damage. *Exp. Neurol.* **166**(2), 266–275.

61. Goldberg, M. A., Dunning, S. P., and Bunn, H. F. (1988) Regulation of the erythropoietin gene: evidence that the oxygen sensor is a heme protein. *Science* **242**(4884), 1412–1415.

62. Srinivas, V., Zhu, X., Salceda, S., Nakamura, R., and Caro, J. (1998) Hypoxia-inducible factor 1alpha (HIF-1alpha) is a non-heme iron protein. Implications for oxygen sensing. *J. Biol. Chem.* **273**(29), 18,019–18,022.

63. Blanchard, K. L., Acquaviva, A. M., Galson, D. L., and Bunn, H. F. (1992) Hypoxic induction of the human erythropoietin gene: cooperation between the promoter and enhancer, each of which contains steroid receptor response elements. *Mol. Cell Biol.* **12**(12), 5373–5385.

64. Guillemin, K. and Krasnow, M. A. (1997) The hypoxic response: huffing and HIFing. *Cell* **89**(1), 9–12.

65. Halterman, M. and Federoff, H. (1999) HIF-1α cooperates with p53 to promote hypoxia induced neuronal death, in *Keystone Symposia—Apoptosis and Programmed Cell Death*, Keystone Symposia, Breckenridge, CO, p. 53.

66. Van de Water, T. R., Staecker, H., Halterman, M. W., and Federoff, H. J. (1999) Gene therapy in the inner ear. Mechanisms and clinical implications. *Ann. N.Y. Acad. Sci.* **884**, 345–360.

67. Ard, M. D., Morest, D. K., and Hauger, S. H. (1985) Trophic interactions between the cochleovestibular ganglion of the chick embryo and its synaptic targets in culture. *Neuroscience* **16**(1), 151–170.

68. Koitchev, K., Aran, J. M., Ivanov, E., and Cazals, Y. (1986) Progressive degeneration in the cochlear nucleus after chemical destruction of the cochlea. *Acta Otolaryngol.* **102**(1–2), 31–39.

69. Pfingst, B. E. and Sutton, D. (1983) Relation of cochlear implant function to histopathology in monkeys. *Ann. N.Y. Acad. Sci.* **405**, 224–239.

70. Spoendlin, H. (1971) Primary structural changes in the organ of Corti after acoustic overstimulation. *Acta Otolaryngol.* **71**(2), 166–176.

71. Webster, M. and Webster. D. B. (1981) Spiral ganglion neuron loss following organ of Corti loss: a quantitative study. *Brain Res.* **212**(1), 17–30.
72. Ylikoski, J., Wersall, J., and Bjorkroth, B. (1974) Degeneration of neural elements in the cochlea of the guinea-pig after damage to the organ of corti by ototoxic antibiotics. *Acta Otolaryngol. Suppl.* **326**, 23–41.
73. Zhou, X. N. and Van de Water, T. R. (1987) The effect of target tissues on survival and differentiation of mammalian statoacoustic ganglion neurons in organ culture. *Acta Otolaryngol.* **104**(1–2), 90–98.
74. Lefebvre, P. P., Van de Water, T. R., Staecker, H., et al. (1992) Nerve growth factor stimulates neurite regeneration but not survival of adult auditory neurons in vitro. *Acta Otolaryngol.* **112**(2), 288–293.
75. Ylikoski, J., Pirvola, U., and Happola, O. (1993) Characterization of the vestibular and spiral ganglion cell somata of the rat by distribution of neurofilament proteins. *Acta Otolaryngol. Suppl.* **503,** 121–126.
76. Ernfors, P., Lee, K. F., and Jaenisch, R. (1994) Target derived and putative local actions of neurotrophins in the peripheral nervous system. *Prog. Brain Res.* **103,** 43–54.
77. Schecterson, L. C. and Bothwell, M. (1994) Neurotrophin and neurotrophin receptor mRNA expression in developing inner ear. *Hear. Res.* **73**(1), 92–100.
78. Chen, X., Frisina, R. D., Bowers, W. J., Frisina, D. R., and Federoff, H. J. (2001) Hsv amplicon-mediated neurotrophin-3 expression protects murine spiral ganglion neurons from cisplatin-induced damage. *Mol. Ther.* **3**(6), 958–963.
79. Hartshorn, D. O., Miller, J. M., and Altschuler, R. A. (1991) Protective effect of electrical stimulation in the deafened guinea pig cochlea. *Otolaryngol. Head Neck Surg.* **104**(3), 311–319.
80. Michelson, R. P. and Schindler, R. A. (1983) Surgical approach for insertion of multichannel electrodes into the scala tympani. *Ann. N.Y. Acad. Sci.* **405,** 343–347.
81. Staecker, H., Gabaizadeh, R., Federoff, H., and Van De Water, T. R. (1998) Brain-derived neurotrophic factor gene therapy prevents spiral ganglion degeneration after hair cell loss. *Otolaryngol. Head Neck Surg.* **119**(1), 7–13.
82. Carenza, L., Villani, C., Framarino dei Malatesta, M. L., et al. (1986) Peripheral neuropathy and ototoxicity of dichlorodiamineplatinum: instrumental evaluation. Preliminary results. *Gynecol. Oncol.* **25**(2), 244–249.
83. Fleischman, R. W., Stadnicki, S. W., Ethier, M. F., and Schaeppi, U. (1975) Ototoxicity of *cis*-dichlorodiammine platinum (II) in the guinea pig. *Toxicol. Appl. Pharmacol.* **33**(2), 320–332.
84. Nakai, Y., Konishi, K., Chang, K. C., et al. (1982) Ototoxicity of the anticancer drug cisplatin. An experimental study. *Acta Otolaryngol.* **93**(3–4), 227–232.
85. Rybak, L. P. (1981) *Cis*-platinum associated hearing loss. *J. Laryngol. Otol.* **95**(7), 745–747.
86. Cavaletti, G. and Tredici, G. (1996) Evaluation of cisplatin neuroprotection by NT-3. *Ann. Neurol.* **39**(6), 827.
87. Zheng, J. L., Stewart, R. R., and Gao, W. Q. (1995) Neurotrophin-4/5 enhances survival of cultured spiral ganglion neurons and protects them from cisplatin neurotoxicity. *J. Neurosci.* **15**(7 Pt 2), 5079–5087.

88. Han, J. J., Mhatre, A. N., Wareing, M., et al. (1999) Transgene expression in the guinea pig cochlea mediated by a lentivirus-derived gene transfer vector. *Hum. Gene Ther.* **10**(11), 1867–1873.

89. Stover, T., Yagi, M., and Raphael, Y. (1999) Cochlear gene transfer: round window versus cochleostomy inoculation. *Hear. Res.* **136**(1–2), 124–130.

90. Bowers, W. J., Howard, D. F., Brooks, A. I., Halterman, M. W., and Federoff, H. J. (2001) Expression of vhs and VP16 during HSV-1 helper virus-free amplicon packaging enhances titers. *Gene Ther.* **8**(2), 111–120.

91. Bowers, W. J., Howard, D. F., and Federoff, H. J. (2000) Discordance between expression and genome transfer titering of HSV amplicon vectors: recommendation for standardized enumeration. *Mol. Ther.* **1**(3), 294–299.

92. Stavropoulos, T. A. and Strathdee, C. A. (1998) An enhanced packaging system for helper-dependent herpes simplex virus vectors. *J. Virol.* **72**(9), 7137–7143.

93. Sena-Esteves, M., Saeki, Y., Fraefel, C., and Breakefield, X. O. (2000) HSV-1 amplicon vectors—simplicity and versatility. *Mol. Ther.* **2**(1), 9–15.

94. Saeki, Y., Fraefel, C., Ichikawa, T., Breakefield, X. O., and Chiocca, E. A. (2001) Improved helper virus-free packaging system for HSV amplicon vectors using an ICP27-deleted, oversized HSV-1 DNA in a bacterial artificial chromosome. *Mol. Ther.* **3**(4), 591–601.

95. Brooks, A. I., Cory-Slechta, D. A., Bowers, W. J., Murg, S. L., and Federoff, H. J. (2000) Enhanced learning in mice parallels vector-mediated nerve growth factor expression in hippocampus. *Hum. Gene Ther.* **11**(17), 2341–2352.

96. Brooks, A. I., Cory-Slechta, D. A., and Federoff, H. J. (2000) Gene-experience interaction alters the cholinergic septohippocampal pathway of mice. *Proc. Natl. Acad. Sci. USA* **97**(24), 13,378–13,383.

97. Brooks, A. I., Halterman, M. W., Chadwick, C.A., et al. (1998) Reproducible and efficient murine CNS gene delivery using a microprocessor-controlled injector. *J. Neurosci. Meth.* **80**(2), 137–147.

Microinjections
to Study the Specific Role
of Proapoptotic Proteins in Neurons

Yan Zhang and Andréa C. LeBlanc

1. Introduction

To manage the instrument successfully, delicacy of touch and a great
deal of patience are required; but it is only with the latter, combined
with perseverance, energy, and close observations that scientific facts
have, or ever will be established.

Dr. H. D. Schmidt, 1895

Since the first description of microsurgery on a single cell in 1835
by Dujardin (reviewed in ref. *1*) microinjection has become a significant tool to physically deliver chemicals, proteins, peptides, nucleic
acids, nuclei, even cells into certain compartments of a single cell. In
the early years, microinjection of embryos was used in the study of
developmental patterns of both invertebrates and lower vertebrates.

In microinjections, the injected material is the only independent
variable in the experiment and, as a result, all effects observed
are attributable to the injected material. Thus, the role of the injected
material in neurons is efficiently dissected out. The injection dosage is precisely controlled. Microinjections can selectively deliver
the material into a certain subcellular compartment, such as the
cytoplasm or the nucleus. One can discriminately inject a certain
group of cells in one culture dish, while the uninjected cells serve
as built-in controls. Finally, injections can be done on cells at a

From: *Neuromethods, Vol. 37: Apoptosis Techniques and Protocols, 2nd Ed.*
Edited by: A. C. LeBlanc @ Humana Press, Inc., Totowa, NJ

particular stage of apoptosis, for example, before or after commitment to apoptosis or synaptic disruption.

However, like every technique, microinjection has its limitations. First, since injection is a physical stress to cells in addition to the experimental treatments, a vehicle control is absolutely required to ensure that the injection itself does not significantly affect cell viability. Second, positive controls and immunohistochemical staining to detect the injected protein or peptide are recommended. Third, microinjection is most powerful to deliver membrane-impermeable chemicals, cDNAs, cytosolic and nucleoplasmic proteins or peptides, but not membrane proteins or ion channels, although this can be achieved by injecting the cDNAs of these membrane proteins. Fourth, while injecting protein or peptide, the injected neuron or cell must be identified in the culture by the co-injection of a membrane-impermeable marker dye such as dextran Texas red (DTR) or the protein conjugated to a fluorescent dye. In dividing cell lines, the dextran Texas Red is diluted with cell division, thereby resulting in the loss of the identity of the injected cell (reviewed in ref. 2). This can be avoided in cDNA microinjections co-expressing a green fluorescent protein (GFP) marker to ensure the continuous detection of the injected cell. Fifth, microinjection is used for single-cell analysis. It cannot be used if large amounts of injected cells are required for biochemical analysis, since injecting huge numbers of cells is extremely time-consuming and labor-intensive. This limitation prevents analysis of gene expression levels of the injected cDNA and does not allow collection of genetic material except when using single-cell analysis *(3)*. Lastly, the system and equipment are costly and the technique requires manual skills that may be challenging for beginners.

However, for certain purposes, the advantages of microinjection by far overweigh its limitations, since it allows quantitative and subcellular delivery of proteins and peptides into mammalian cells, especially cells that are difficult to transfect or infect, as are cultured primary neurons *(see* detailed discussion under Subheading 2.3.). Using this technique, we have successfully studied the roles of recombinant active caspases *(4)*, Bax cDNA expression *(4a)*, and intracellular amyloid β (Aβ) peptides in human neuronal apoptosis (manuscript in preparation). The technical details and protocols of microinjection of recombinant proteins, peptides, and cDNAs are discussed below.

2. Microinjection of Molecules into Cells

2.1. Microinjection as a Tool to Deliver Proteins into a Single Neuron

Protein–protein interactions and protease activation in the case of the Bcl-2 family members and caspases, respectively, represent two mechanisms regulating neuronal survival or cell death. The role of these proteins in apoptosis is often detected by studying the translocation or posttranslational modifications of Bcl-2 or Bax proteins or the activation of caspases in insulted neurons. Because the apoptotic insult can induce a number of other cell death mechanisms that could go undetected, the microinjection of these proteins in neurons addresses the direct role of these specific proteins in an otherwise unaltered neuron. Proteins can be directly delivered into a single cell by either electropermeabilization (electroporation) (5, reviewed in ref. 6), or microinjection. Electroporation requires large amounts of protein compared to microinjection and is seldom used on neurons (7). Microinjection has been used to deliver a number of proteins into neuronal cells to study apoptosis. We have injected human cultured primary neurons with recombinant active caspases to determine which caspases are responsible for human neuronal apoptosis (4). In other studies, rat superior cervical ganglia (SCG) cells were injected with the antibodies to Jun and Fos proteins to study their roles in NGF-deprivation-induced apoptosis (8). Cytochrome-*c* was injected into hybrid motor neuron 1 cells (9), sympathetic neurons (10,11), and rat SCG neurons (12). Similarly, Bid and AIF were injected into Rat-1 fibroblasts to study the effects of these proteins on the permeability transition pore complex and understand their mechanisms of action in apoptosis (13,14).

2.2. Microinjection as a Tool to Deliver Peptides into Single Neurons

Microinjection of intracellular peptides in neurons can also allow the study of suspect peptides resulting from abnormal protein metabolism such as the Alzheimer's disease amyloid β peptide (manuscript in preparation). Alternative methods to deliver peptides include delivery by a carrier peptide that is linked to the peptide of interest. The link, usually made by disulfide bonds, is cleaved intracellularly and the peptide of interest released in the cell (reviewed in ref. *15*). However, this method is limited in that it is impossible to control the timing and the exact amount of peptide released in the cell.

2.3. Microinjection as a Tool to Deliver cDNAs into Single Neurons

While microinjection of cDNA in cells also lacks fine control on the level of gene expression, it can be somewhat controlled by injecting varying doses of cDNA. Various methods other than microinjections can transduce genes into eukaryotic cells. Transfection of cDNAs into mammalian cells is routinely done using calcium phosphate, DEAE-dextran, or liposome-mediated transfection and electroporation. In calcium phosphate and DEAE-dextran transfections, chemicals are used to attach foreign DNAs to the cell surface, which allows endocytosis of the foreign DNAs into cells. Electroporation results in pore formation on cell membrane by an electric field. Since a cuvet is needed during the electric shock, this method is more suitable for cells in suspension rather than for attached differentiated neurons. In general, liposomes contain cationic and neutral lipids, which bind to negatively charged DNA phosphate groups. Liposomes possibly anchor on the cell membrane and deliver DNAs into cells *(16)*. Unfortunately, primary neurons are less resistant to transfections by these methods than most cell lines. Furthermore, selection of the transfected cells using antibiotics leads to cell death of the nontransfected neurons, a state that is undesirable since it will likely damage the neuronal network necessary for the survival of the cultured neurons.

Viral infections have been widely used to introduce cDNAs into neuronal cells. The gene of interest is inserted into a replication-defective viral construct encapsulated within a virus envelope that increases the efficiency of gene transfer. This method is able to achieve high levels of transfection *(17)*. However, often these infections are toxic to some extent to neurons. In addition, choosing the proper virus, constructing a recombinant virus, and packaging the virus involve considerable effort and time *(17)*. If the expressed protein is toxic to the packaging cells, an inducible system is required.

Comparatively, microinjection of the cDNA in the neuron cannot be used to generate large amounts of transfected cells and is therefore limited to observations at the single-cell level. However, it is an excellent method when testing the effects of one or more proteins on a defined function such as apoptosis, since more than one cDNA can be microinjected together in primary neurons. Injection of the cDNA into cells is less disruptive to the cell than the transfection or infection techniques, is a rapid and efficient manner to deliver known quantities of cDNA, and overcomes a problem of background apoptosis often observed with viral or liposome transfections. The injection of antisense oligonucleotides can also avoid DNA delivery problems from endocytosis and result in the rapid translocation of the antisense oligonucleotide to the nucleus (15,18).

3. Microinjection of Neurons and Neuronal Cell Lines

Virtually all cells, including cell lines and primary cells in culture, can be microinjected. Suspension cells are more challenging than attached cells, since a holding pipet with a negative pressure is required to anchor the injected cell. However, for attached cells such as primary neurons or neuroblastoma cell lines, only one injection pipet is required.

Neuronal cell lines are commonly used for the study of apoptosis since they are relatively easy to culture and transfect or infect. However, microinjections can be useful for specific purposes as discussed above. We have injected several cell lines such as human teratocarcinoma NT2, neuroblastoma M17 cells, and LAN1 cells. These cells are easier to inject because they have larger cell volume and cytoplasm than the primary neurons (unpublished results). Other neuronal cell lines such as rat pheochromocytoma PC12 cells have also been microinjected (19). The problem with injecting cell lines is that the injected recombinant protein, peptide, or oligonucleotide will be diluted and eventually lost during cell division. In addition, since apoptotic mechanisms are often insult-, cell type-, and occasionally species-specific, the study of apoptotic processes using cell lines may not reflect exactly that of postmitotic neurons. Primary cultured neurons are, therefore, most suitable as a culture model to study the mechanism of neuronal apoptosis. However, the results of these experiments cannot be extended directly to the in-vivo situation without verification of the mechanism in an in-vivo model. In our laboratory, we use primary cultures of human neurons (20). These cells are quite resistant to microinjections compared to human primary astrocytes and several neuronal cell lines that we have studied (Table 1). A survival rate of nearly 90% for

Table 1
Suggested Injection Parameters for Different Human Cells

	Compensation pressure (hPa)	Injection pressure (hPa)	Injection time (s)	Injection volume (pL)	Cell viability 16 d after injections
Neurons	40–50	100	0.1	25	90%
Astrocytes	20–30	50	0.1	8–10	50%
M17	20–30	50	0.1	8–10	50%
NT2	20–30	50	0.1	8–10	50%
LAN1	20–30	50	0.1	8–10	50%

human neurons, and 50% for human astrocytes, teratocarcinoma cells NT2, and neuroblastoma M17 and LAN1 cells, can be achieved by an experienced researcher using the procedure described below (4). However, the small cytosolic compartment of primary neurons makes microinjections very challenging to a beginner, and can only be achieved with practice.

4. Microinjection Equipment

4.1. Microscope and Light Source

An inverted microscope with adjustable magnification of at least 200X is required. For experiments involving nuclear injection, a magnification of 400X may be necessary. A good view of depth and an inverted field are also recommended since the noninverted microscope limits largely the working distance between the lens and objectives (21). We currently use the IMT-2 Research Microscope from Olympus Optical Co. (Lake Success, NY) at 200X magnification for cytosolic injection. The pseudo-three-dimensional image provided by the Nomarski or Hoffman optical system can help the intricate positioning during nuclear injection (21). The Nomarski system, however, is designed for an optical path through glass only, hence it is not suitable to obtain a correct image from a plastic specimen container or cover slip (22). Typically, the illumination from microscopes, such as halogen bulbs, is good enough for cytosolic injections (22). If nuclear injection is desired, a fiberoptic light source is necessary to visualize the boundary of the nucleus (21,22).

4.2. Micromanipulator

A micromanipulator that allows fine and smooth three-dimensional movements is needed to position the injection needle. A micromanipulator that reduces most of the movements and vibrations from hand and arm is preferred (22). In addition, the micromanipulator should be placed on a solid, stable, and flat table to reduce vibrations from the environment as much as possible. The MIS-5000 series Microinjection Micromanipulator from Burleigh Instruments (Fishers, NY) is used in our study (4) since it is characterized by the easy and fine control of not only horizontal but also vertical movements by its joystick (MIS-503) to achieve stable and high-resolution positioning. As showed in Fig. 1A, the micromanipulator (MIS-520) is mounted on the microscope. The injection pipet holder is mounted on the micromanipulator. A Petri dish serving as an injection chamber (*see* details under Subheading 4.5.) is placed on the platform of the microscope. The injection needle is placed on the top, at an angle of 30–45° relative to the cells (Fig. 1B). A number of other systems are available and perfectly appropriate for microinjections.

4.3. Microinjector

A microinjector controls the compensation pressure, injection pressure, and injection time. For most microinjectors, injection is controlled by air pressure using compressed nitrogen or air (21). Compensation pressure is required to overcome capillary forces caused by the fairly fine tip of the injection needle. Due to capillary forces, liquid from the culture dish tends to flow into the injection needle and, thereafter, dilute the injection solution inside of the needle. To prevent this, the compensation pressure is set to push the solution inside of the needle to the tip and ready to be injected (Fig. 2). A value that ensures a continuous trace amount of discharge out of the needle tip should be determined before experiments by using a visible dye such as fast green (Sigma) to detect the trace discharge. The injection pressure and compensation pressure, together with injection time, decide the volume of injection and therefore determine the applied dose into every cell (*see* Subheading 5.3.2. for details).

The Transjector 5246 from Eppendorf (Hamburg, Germany) is used in our study of cytosolic injection of single cells (4). This device has a total of four output channels: two are available for

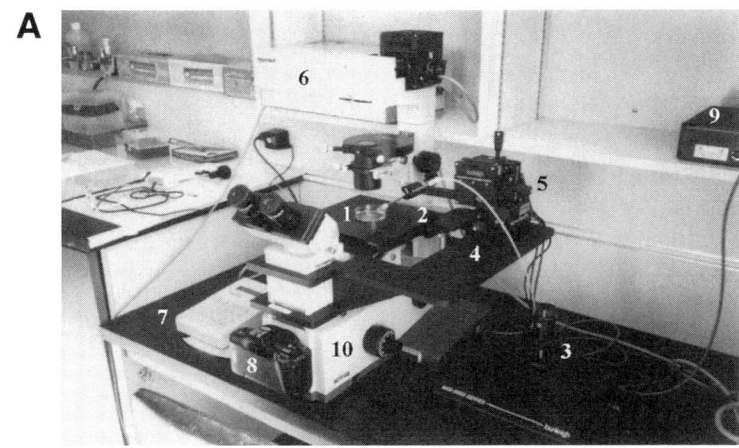

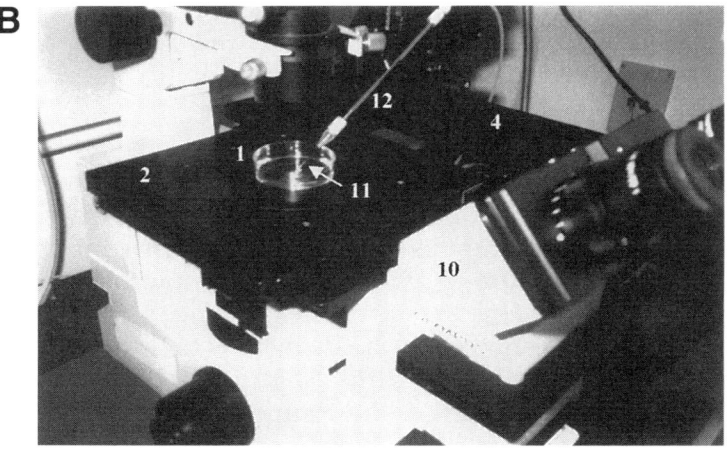

Fig. 1. Microinjection equipment. (**A**) A picture of a typical microinjection work station, including a microscope, a camera, a micromanupulator with a joystick, a microinjector, the area of injection, and an injection needle mounted on the micromanipulator. (**B**) A close-up picture of injection area. A Petri dish containing culture medium is used to hold the cover slip to be injected. The injection needle is inserted into the chamber on the top of the neurons at around 30–45°. 1, injection chamber; 2, microscope platform; 3, joystick; 4, platform for micromanipulator attached to the microscope; 5, micromanipulator; 6, microinjector; 7, hand control pedal; 8, camera; 9, power supply; 10, microscope; 11, injection needle; 12, needle holder.

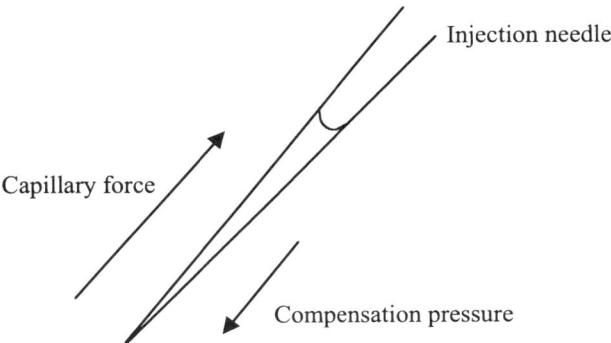

Fig. 2. Compensation pressure is required against the capillary force from the injection needle. A schematic diagram showing that the capillary force resulting from the tip of the injection needle is compensated by the compensation pressure from the microinjector.

microinjection; the other two are used to hold cells by negative pressure if necessary. This system delivers constant volumes and has helpful functions such as hold, fill in, and clear tip. Furthermore, injections can also be performed by a foot pedal. Other systems are also recommended (21,22).

4.4. Injection Needle, Glass Pipet, and Puller

The preparation of the injection needle is critical for a successful microinjection. In our experiments, the needles are pulled from the thin-walled MTW100F-4 borosilicate glass capillaries with microfilament (World Precision Instruments, Sarasota, FL). These capillaries have 1.0-mm outer diameter and 0.5-mm inner diameter. The microfilament ensures that the injection solution flows easily to the needle tip. Also, glass capillaries from Sutter Instrument (Novato, CA) are commonly used (1,21).

Injection needles are pulled by the Flaming/Brown P-87 Micropipetter Puller from Sutter Instrument (Novato, CA) (Fig. 3A). The puller electrically heats up the glass capillary with a heating element. Then a horizontal linear force pulls the heated glass apart to produce two tapered needles (22). A gas jet is underneath the heating element. During the end stage of pulling, gas, usually nitrogen or compressed air, is blown by the gas jet vertically to break the glass. Therefore, pulling force, pulling speed, glass tempera-

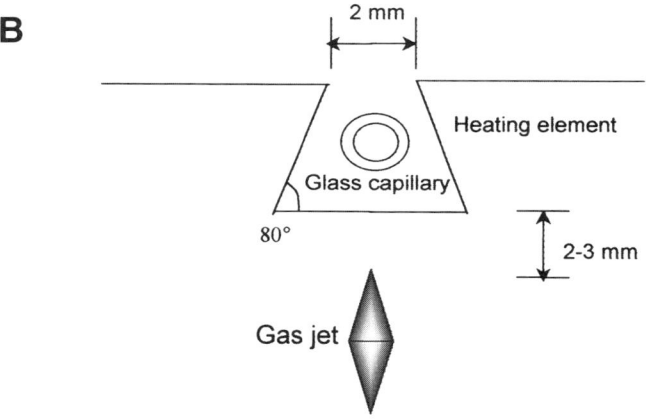

Fig. 3. Needle puller. (**A**) A picture of the Flaming/Brown P-87 Micro-pipetter Puller (Sutter Instrument). A glass capillary is mounted on the puller. When the heating element heats up the glass, a horizontal pulling force draws the glass capillary apart as indicated by the arrows. (**B**) A schematic diagram showing the shape of the heating element and the relative position of glass capillary, heating element, and gas jet of the puller.

ture, and gas blowing time are the critical parameters, for needle shape and tip diameter. Preliminary experiments are needed to establish these parameters, which can vary with the type of glass capillaries, air humidity, room temperature, and the shape of the heating element inside of the puller *(22)*.

To establish the parameters of the puller, the shape and location of the heating element are first adjusted. It is recommended by the manufacturer (World Precision Instruments, Sarasota, FL) to bend the heating element walls to 80° for both sides (Fig. 3B) and keep the upper opening two walls at 2 mm apart. The heating element should be placed 2–3 mm away from the gas jet of the puller. The glass capillary is placed in the center of the two walls of the heating element (Fig. 3B). Then the proper glass temperature is determined. Glass melting temperature typically ranges from 300 to 700°C depending on the shape and age of the heating element, the distance between the heating element walls to the glass capillary, and the glass type of the capillary. In general, higher temperature results in greater plasticity, i.e., finer-tipped needles with a longer tapered area. However, very high temperatures result in long wisps of glass during pulling *(22)* and also risk damaging or aging the heating element.

The pulling force and speed are also critical for needle tip shape. In general, greater pulling force and speed result in finer and sharper tips. A multiple-step pulling is useful for a fine-tuned tip shape. During multiple-step pulling, the movement of capillary in each step is restricted during pulling. Therefore, the tip gets finer in each step.

The needles used for cytosolic microinjection in our study ideally are gradually tapered with the distance between the shaft and the tip as 0.4–0.6 cm and the approximate tip diameter as 0.5 µm measured by a calibrated reticule within the ocular of a stereomicroscope. Once pulled, needles are stored in a micropipet rack with a cover at room temperature in a dry and clean area. Commercially available ready-to-use pipets such as the Femtotips (Effendorf, Hamburg, Germany) are also available.

4.5. Microinjection Chambers

A Petri dish with culture medium is used as a chamber for injection in our study. Cells are cultured on ACLAR (33C; 5 mm; Allied Chemical, Minneapolis, MN) or glass cover slips (*see* under Subheading 5.1.), and these are placed into the Petri dish in culture medium during the injection (Fig. 1B). Since the manipulation time is relatively short (around 5 min to inject 100 cells on one cover slip), neurons survive at room temperature and the medium retains its proper pH. For injections requiring longer manipulation duration, a heated platform and CO_2/O_2 chamber may be attached to the injection platform *(21)*.

5. Microinjection of Recombinant
Active Caspases in Primary Neurons

Here, we describe the microinjection of recombinant active caspases into primary cultures of human neurons and astrocytes as an example for a protein injection. It has been well defined that caspases play an important role in neuronal apoptosis and are possibly crucial to neuronal loss in neurodegenerative diseases, such as Alzheimer's disease (reviewed in refs. *23* and *24*). Our initial studies in the identification of caspases in human neuronal cell death had unexpectedly shown that caspase-6, but not caspase-3, was activated in serum deprived primary cultures of human neurons *(25)*. To address whether caspase-6 was directly responsible for cell death or whether the activation of caspase-6 was only one of many possible events that occurs during neuronal apoptosis, we choose to microinject the recombinant active caspases in these neurons in the absence of any other insults. The microinjections determined that caspase-6, but not caspase-3, -7, or –8, induced human neuronal apoptosis *(4)*. Furthermore, the proapoptotic activity of caspase-6 did not require caspase-3 activity since it was not inhibited by caspase-3 or 7 inhibitor Z-DEVD-fmk, indicating that caspase-6 works as an effector caspase in these neurons. As mentioned under Subheading 2.1., we delivered a precise dose of recombinant caspases into a particular cellular compartment by microinjection. We were able to deliver different concentrations of the caspases to assess the physiological relevance of the active caspase into the neuronal cytosol. In addition, immunostaining using antibodies to the injected caspase-6 was used to determine the lifespan of the injected protein relative to the apoptotic profile. For example, by combining microinjection with immunohistochemistry, we showed that the injected active caspase-6 persisted in neurons for less than 48 h, whereas apoptotic TUNEL-positive neurons continuously underwent cell death between 2–6 d after the injection. These results indicated that active caspase-6 initiates certain irreversible pathways that eventually lead to cell death and that only a transient activation of the caspase is necessary for the commitment of neurons to apoptosis.

5.1. Cell Cultures

5.1.1. Primary Cultures of Human Neuron and Astrocyte

Primary cultures of human neurons and astrocytes are prepared from 12- to 16-wk-old fetal brains following the protocol described before *(20)*. In brief, meninges and blood vessels from fresh brain tissues are removed, the tissue is sliced in small pieces with scalpels, dissociated with 0.25% trypsin at 37°C, which is then inactivated by 10% decomplemented fetal bovine serum (HyClone). The dissociated tissue is then treated with 0.1-mg/mL DNase I and triturated to make a homogeneous mixture of cells. After filtering the mixture through 130- and 70-μm filters, the flow-through is centrifuged to pellet the cells. The pellet is then washed once in phosphate-buffered saline (PBS) and once in minimal essential media (MEM) with Earle's balanced salt solution containing 0.225% sodium bicarbonate, 1 mM sodium pyruvate, 2 mM L-glutamine, 0.1% dextrose, 1X antibiotic penicillin-streptomycin (all products are from Gibco) with 5% decomplemented fetal bovine serum. Cells are plated on poly-L-lysine-coated ACLAR or poly-L-lysine/laminin coated glass cover slips at the density of 3×10^6 cells/mL. Usually, the plated neurons attach to the cover slip within 24 h. In general, the cultures contain 90–95% neurons and 5–10% astrocytes *(20)*. Microinjection is performed 10 d after plating for both neurons and astrocytes *(4)*.

5.1.2. M17, NT2, and LAN1 Cell Line Cultures

Human neuroblastoma M17 cells (obtained from Dr. June Biedler, MSKCC Laboratory of Cellular and Biochemical Genetics, New York, NY) are cultured on ACLAR cover slips at the density of 1×10^6 cells/mL in Opti-Mem containing 5% fetal bovine serum. Human teratocarcinoma NT2 (Stratagene, La Jolla, CA) and neuroblastoma LAN1 cells (obtained from Dr. Culp, Case Western Reserve University, Cleveland, OH) are cultured on ACLAR or glass cover slips at a density of 1×10^6 cells/mL in DMEM containing 10% fetal bovine serum. Microinjections are performed when the cells reach 70–80% confluence on the cover slips.

5.2. Preparation of Proteins and Fluorescent Dye Marker for Microinjections

All solutions for injections should be sterilized. Solutions containing detergents (NP40, Brij3 or 5, Triton X-100, etc.), glycerol above 20%, bacteriostatics such as sodium azide and salt solutions above 400 mM are not suitable for microinjection. Cells are unable to balance intracellular pH against strong buffers. The cells should be microinjected with a buffer that induces the least perturbation. Therefore, always check the possible toxicity of the injection solution components in the absence of recombinant active caspases.

5.2.1. Dextran Texas Red

Dextran Texas red (DTR) (3000 MW; Cedarlane Laboratories, Hornby, Ontario, Canada) is injected with the recombinant caspases as a fluorescent marker to recognize the injected cells. The DTR is dissolved at 100 mg/mL in phosphate-buffered saline (PBS; 140 mM NaCl, 2.7 mM KCl, 4.3 mM Na_2PO_4, 1.5 mM KH_2PO_4, pH 7.4) and stored at −20°C. Mix the DTR with the proteins to be injected immediately before the injection. Avoid frequently freezing and thawing of the stock solution.

5.2.2. Preparation of Recombinant Active Caspases

Recombinant caspase-3, -6, -7, and -8 (Pharmingen, San Diego, CA) are prepared in the caspase active buffer containing 20 mM piperazine-N,N'-*bis*-(2-ethanesulfonic acid) (PIPES), 100 mM NaCl, 10 mM dithiothreitol (DTT), 1 mM EDTA, 0.1% 3-[(3-chol-amidopropyl)-dimethylammonio]-2-hydroxy-1-propanesulfonic acid (CHAPS), 10% sucrose, pH 7.2 (*25a*). The stock solutions of caspases are prepared in the caspase active buffer at 100 µg/mL and stored at −20°C. The caspases are diluted at the required concentration with DTR immediately before use.

5.3. Microinjection of Recombinant Active Caspases into the Cytosol of Neurons

5.3.1. Establishing the Microinjection Parameters

Before starting to microinject the caspases, the compensation pressure, injection pressure, and injection time of the microinjector need to be established. The suggested parameters for some cell types are listed in Table 1. However, they vary greatly with the type of equipment used, cell type, and needle diameter and shape. Different cells have different resistance to microinjections as shown in Table 1. In general, primary cultures of human neurons survive microinjections more easily than human astrocytes, teratocarcinoma NT2 cells, and several neuroblastoma cell lines such as M17, NT2, and LAN1 cells.

1. Set the injection pressure in the range of 50–100 hPa. Set the compensation pressure at approximately one-half of the injection pressure. Set the injection time as 0.1–0.5 s.
2. Fill one needle with the injection solution using the Microloader loading tip (Eppendorf, Hamburg, Germany) and avoid any air bubbles inside the needle. Use a visible dye such as fast green to ensure that a trace amount of solution is continuously released from the needle tip under the compensation pressure.
3. Place the needle in the injection holder. Be very careful not to bump the needle when proceeding to step 3.
4. Carefully take out one cover slip and place it into a Petri dish with culture medium. Place the Petri dish onto the stage of the microscope. Adjust focusing until the structure of cytosolic and nuclear areas appears.
5. Select an area containing morphologically healthy and well-connected neurons. It is preferable to inject many cells in one area. This will facilitate later detection of the injected neurons.
6. Position the needle on the top of the cell using the manual micromanipulator and joystick. For cytosolic injection of neurons (Fig. 4A), place the needle just on the top of the cell surface, without piercing the cell membrane. Carefully lower the tip down by slowly moving the vertical control of the joystick to let the tip penetrate into the cytosolic area of the cell (Fig. 4B). This step is critical for a successful injection and cell survival from the injection. Great patience and effort are needed for beginners to achieve successful and consistent injection.
7. Press the "inject" button of the microinjector to finish the injection. Carefully remove the needle from the cell with the joystick.
8. Inject some cells and carefully observe the morphological change and cell behavior immediately after injection. Typically, inflation or explosion of the cell during injection will show you that the injection pressure is too high. If this happens, lower the pressure by increments of 5 hPa until the phenomenon is eliminated.
9. After injection, incubate the cells for at least 96 h or longer, and check cell survival to ensure that the injection itself does not affect cell viability.
10. If cells survive, you are almost ready to microinject. First, perform the following preliminary experiments.

A

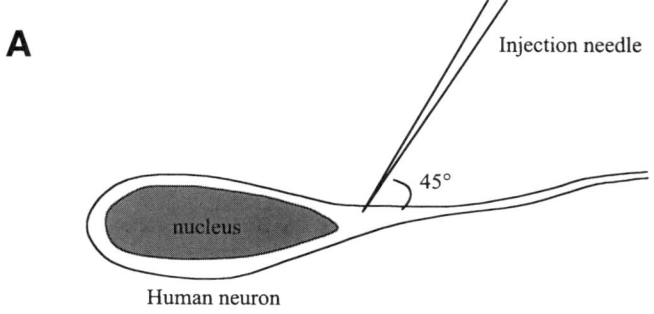

B

Step 1. Position the injection needle on the surface of the cell

Step 2. Lower down the injection needle to penetrate cell membrane

Fig. 4. Microinjections into neurons. (**A**) A schematic diagram showing the position of cytosolic injection needle insert of a neuron. (**B**) Two steps to position the injection needle on the surface the cell, and inserting the needle into cell membrane.

5.3.2. Preliminary Control Experiments
Before Microinjecting the Active Caspase

1. Determine the injected volume. The injection volume can be determined by loading the needle with 0.1–0.2 µL and simply injecting the solution in a cell. Shift cells when the cell bursts. Count the number of ejections required to finish the solution (this can be counted automatically by the transjector), and then the ejection volume can be calculated by dividing the volume loaded in the pipette by the number of ejections. Based on the ejection volume and solution concentration, calculate the actual amount of material delivered for each shot of injection. Cytosolic volumes vary from different cell types, and this should

be taken into account when calculating the dosage of injection in each cell. Typically, we inject 25 pL under our conditions.

2. Optimize the concentration of DTR. Before you start injecting the protein of interest, the concentration of DTR used for microinjection needs to be determined by injecting various doses of DTR (in PBS), ranging from 10 to 1000 µg/mL, into neurons, and counting the number of DTR-positive cells that remain after injection. As shown in Fig. 5A, injection of DTR at 100–300 µg/mL results in the maximal detection of positive neurons. A dose of 1000 µg/mL appears to be toxic. Therefore, we opted to use DTR at a final concentration of 100–200 µg/mL for our experiments but increase this concentration to 400 µg/mL in dividing cells.

3. Determine cellular viability after the injection. Cell viability after injection needs to be assessed. Inject neurons with 200-µg/mL DTR in PBS. Measure cell viability using terminal deoxynucleotidyl transferase-mediated biotinylated UTP nick end labeling (TUNEL) staining after various time of injection ranging from 0 to 20 d. As shown in Fig. 5B, injection of 200-µg/mL DTR does not affect neuronal survival up to 16 d after the injection.

4. Determine the toxicity of the injected buffers. The active caspase buffer contains a number of chemicals that could be toxic to neurons. Before you start, test the toxicity of the caspase buffer in the presence of the predetermined optimal concentration of DTR in neurons. We compared the buffer with PBS. The caspase active buffer itself does not affect cell viability significantly compared with PBS control (Fig. 5C).

5.3.3. Microinjection of the Caspases

1. Prepare the injection solutions by mixing the DTR with buffers or recombinant active caspases at the desired concentration. At first, you will need to establish the conditions of microinjection as described under Subheading 5.3.1. You will also have to establish the optimal concentration of nontoxic doses of DTR required to detect the injected neurons as described under Subheading 5.3.2. Follow steps described above to inject the neurons.

2. Repeat the injection on as many cells as you wish. We typically inject 100 cells/cover slip and inject two cover slips for each preparation of neurons. However, for cells that are less resistant to microinjections, more cells might need to be injected so

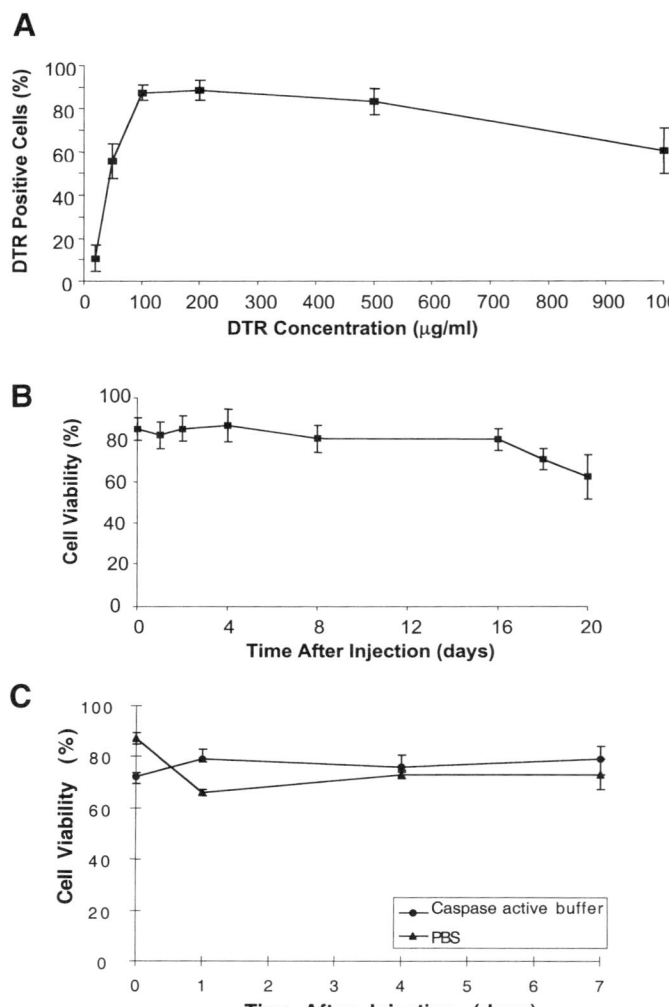

Fig. 5. Preliminary experiments before testing the apoptotic role of proteins by microinjection. (**A**) DTR (in PBS) was injected into human neurons at various concentrations ranging from 10 to 1000 µg/mL. The number of DTR-positive cells was counted 96 h after injection. As shown here, injection of DTR at 100–300 µg/mL results in the maximal detection of microinjected neurons. (**B**) Neurons were injected with 200-µg/mL DTR in PBS. Cell viability was measured by TUNEL staining after various time of injection ranging from 0 to 20 d. The injection of 200-µg/mL DTR does not affect neuronal survival up to 16 d after injection. (**C**) Effect of caspase active buffer on cell viability tested against PBS. After 7 d of injection, the caspase active buffer does not affect cell viability significantly.

you end up with 100 cells to analyze. It is easier to detect cells that are injected in the vicinity of each other. You can repeat multiple injections in different sites of the same cover slip.

3. After microinjections, cells are returned to the incubator and incubated for a desired period of time. In our primary human neuron cultures, we verify apoptosis at 1, 2, 4, 7, and 16 d, but a different time scale may be needed with a different cell type. We measure apoptosis by TUNEL staining and morphological shrinkage (*see* Subheading 5.4.), but other methods to detect various stages of apoptosis can also be used. If you have a live fluorescence microscope and your cells detach when dying, you can estimate the loss of DTR-positive cells at each time point in the same injected culture. Once a survival curve is obtained, you can verify the apoptotic cell death with standard methods.

5.3.4. Technical Precautions

1. Theoretically, only soluble proteins can be injected efficiently, although insoluble proteins with small molecular weight and proper conformation that do not clog the injection needle could be injected. Some sticky proteins, large proteins, and strongly charged proteins are not suitable for injection, since they might block the injection needle tips. You can confirm that you injected the protein of interest by performing immunocytochemistry on the injected cells.
2. Injection of the denatured, inactive form of the caspase is also recommended to rule out the possible apoptotic effect of introducing foreign protein. Denaturation of the active caspase can be achieved by boiling the protein for 10 min.
3. Check the active caspase solution frequently to ensure that the protein is still active and functioning. Here, we tested the activity of recombinant caspases used for injection by in-vitro caspase activity assay *(4)*.

5.4. Measurement of Apoptosis

5.4.1. Solutions and Products

1. Fixation solution: 4% paraformaldehyde and 4% sucrose (in PBS, pH 7.4) is prepared immediately before use as described *(26)*.
2. Washing buffer: PBS.
3. Permeabilization solution: 0.1% Triton X-100 (in 0.1% sodium citrate).

4. TUNEL reaction mixture: prepared as described by the manufacturer (Roche Molecular Biochemicals, Hertforshire, UK).
5. Mounting medium for light microscopy: We use Geltol mounting medium for light microscopy, Immunon™ from Shandon, Pittsburg, PA.

5.4.2. Controls

1. Negative control: Incubate fixed and permeabilized samples in 50-µL/well labeling solution without terminal transferase.
2. Positive control: Incubate fixed and permeabilized samples with micrococcal nuclease or DNase I (1 µg/mL–1 mg/mL, 10 min at room temperature) to induce DNA breaks.

5.4.3. Protocol

Cells are fixed in fresh 4% paraformaldehyde, 4% sucrose for 20 min and permeabilized with Triton X-100. TUNEL is performed using the *in-situ* Cell Death Detection Kit I as described by the manufacturer. The percentage of apoptosis is determined by the ratio of the number of DTR-TUNEL-double-positive cells over the total number of DTR-positive cells *(4)*. Here, we briefly describe the TUNEL staining protocol for the *in-situ* Cell Death Detection Kit I.

1. Aspirate the medium from the microinjected cells and cover the cells with freshly prepared fixation solution for 20 min at room temperature.
2. Rinse the cover slips with PBS by aspirating the fixative, add 500 µL of fresh PBS to cover the cells, and incubate for 5 min at room temperature.
3. Aspirate and repeat the rinse twice. At this point, you can store the fixed cells at 4°C and continue the procedure later.
4. To permeabilize the cells, aspirate the PBS, cover the cells with permeabilization solution, and incubate for 2 min on ice.
5. Rinse the cover slips with PBS three times as before.
6. Air-dry the area around the cells by aspirating with a fine pipet tip.
7. Add the TUNEL reaction mixture on the cells. Since we plate our neurons in a 50-µL drop, 10 µL of TUNEL mix is enough to cover the area of the cells. You may need to adjust the volume depending on your cell culture. Make sure that the cells do not dry.
8. Incubate the cover slips in a humidified chamber for 60 min at 37°C. From this step, avoid leaving the cells in bright lights.
9. Pick up the cover slip from the well and dip 8 times in PBS and 8 times in nanopure sterile distilled water.

10. Blot the excess liquid from the cover slip and mount the cover slip by inverting it on a glass slide covered with a small drop of mounting medium. It is best to leave the slides to settle overnight in the dark before analyzing under a fluorescence microscope.

11. You first should identify the red (DTR-positive) microinjected cells. Once you have identified these, change your filter to determine if these are also green (TUNEL-positive). Calculate the percentage apoptosis by dividing the neurons that are red (DTR-positive) and green (TUNEL) by the total number of injected neurons that have survived (DTR-positive). Check your negative and positive controls carefully to make sure that the TUNEL worked. Carefully assess the morphology of the cells since necrosis can also give TUNEL-positive neurons. In the absence of DNA fragmentation but cell death, monitor up-stream apoptotic events such as caspase activation or Annexin V staining.

6. Microinjection of Peptides into Neurons

The importance of peptides such as Alzheimer's disease amyloid β peptide (Aβ) or the caspase-generated C-terminal amyloid precursor protein fragments *(25,27–30)* in the pathophysiology of neurodegenerative diseases can require testing the toxicity of these peptides inside neurons. Microinjections can also be used to assess the intracellular toxicity of these or other peptides. The microinjection of peptides has the same advantages outlined above for the protein microinjection. Here we describe techniques developed in our laboratory to test the effect of such peptides in neurons.

6.1. Preparation of Peptides

Peptides are bought or synthesized from a reliable source. Peptides can be dissolved in water, PBS or mild buffers at 25 µM or less and stored in aliquots at –20°C. The conformation of the peptide may be critical for toxicity, so it is wise to obtain the physical structure of the peptide to be injected and monitor different batches of peptides carefully.

6.2. Preparation of the Cells

See Subheading 4.1.

6.3. Microinjection of Peptides and Measurement of Apoptosis

The peptides are diluted in PBS at a concentration ranging from 0.1 to 10 n*M* solution with DTR as described above immediately before the microinjection. It is important to use a reverse or scrambled peptide as a control. It is also important to know the structure of the peptide to be injected. As with protein aggregates, it may not be possible to microinject fibrillar peptides, as they may plug the injection needle. It is wise to monitor the ejectate for structure. Avoid freezing and thawing. For details on microinjection and detection of apoptosis follow Subheadings 5.3.1.–5.3.3.

7. Microinjection of cDNAs into Neurons

To explore the function of Bax, a proapoptotic protein, in human primary cultured neurons, we microinjected Bax cDNA construct into the cytosol of neurons. Human bax alpha cDNA was inserted into a eukaryotic expression vector, pCep4β, under the strong CMV promoter to ensure high expression levels of Bax protein.

7.1. Preparation of cDNAs to Be Microinjected

All cDNAs for microinjection have to be purified by column purification with commercially available plasmid purification kits, such as the Plasmid Midi Kit (Qiagen). DNAs from minipreps are not recommended for injection, since they will contain a number of impurities. Avoid using phenol/chloroform extraction methods. Typically, the cDNA constructs are dissolved in water as stock solutions and stored at –20°C. Before use, dilute the cDNA at 30 µg/mL in PBS with 100–200-µg/mL DTR.

7.2. Microinjection and Measurement of Apoptosis

We typically inject 25 pL of a 30-µg/mL Bax cDNA construct with DTR as marker dye. At this concentration, Bax induces apoptosis in 70–80% of the primary human neurons within 12 h of injection. Various doses of cDNA can be tested initially, since the expression level will vary with different promoters. As a control, we inject the empty vector and DTR alone. The microinjections are done as described under Subheading 5.3. The cells are usually incubated for various periods of time depending on the cell type. Apoptosis is measured as described under Subheading 5.4. Under these conditions, 70–80% of neurons die within 24 h *(4a)*.

Acknowledgments

We are grateful to Dr. Cynthia G. Goodyer for providing human tissues, Jennifer Hammond and Megan Blacker for technical support, and Dr. Shuhui Wu and Dr. Jack Kelly at Carleton University (Ottawa, Canada) for their insightful advice to YZ on micromanipulation. This work is supported by the Alzheimer Society of Canada to YZ and the Medical Research Council of Canada (MT-15118) and FRSQ to ALB.

References

1. Malter, H. E. (1992) Early development of micromanipulation, in *Micromanipulation of Human Gametes and Embryos* (Cohen, J., Malter, H. E., Talansky, B. E., and Grifo, J., eds.), Raven Press, New York, pp. 1–24.
2. Wadsworth, P. (1999) Microinjection of mitotic cells. *Meth. Cell Biol.* **61,** 219–231.
3. Kacharmina, J., Crino, P., and Eberwine, J. (1999) Preparation of cDNA from single cells and subcellular regions. *Meth. Enzymol.* **303,** 3–18.
4. Zhang, Y., Goodyer, C., and LeBlanc, A. (2000) Selective and protracted apoptosis in human primary neurons microinjected with active caspase-3, -6, -7, and -8. *J. Neurosci.* **20,** 8384–8389.
4a. Bounhar, Y., Zhang, Y., Goodyer, C. G., and LeBlanc, A. (2001) Prion protein protects human neurons against Bax-mediated apoptosis. *J. Biol. Chem.* **276,** 39,145–39,149.
5. Bobinnec, Y., Khodjakov, A., Mir, L. M., Rieder, C. L., Edde, B., and Bornens, M. (1998) Centriole disassembly in vivo and its effect on centrosome structure and function in vertebrate cells. *J. Cell Biol.* **143,** 1575–1589.
6. Orlowski, S. and Mir, L. M. (1993) Cell electropermeabilization: a new tool for biochemical and pharmacological studies. *Biochim. Biophys. Acta* **1154,** 51–63.
7. Teruel, M. N., Blanpied, T. A., Shen, K., Augustine, G. J., and Meyer, T. (1999) A versatile microporation technique for the transfection of cultured CNS neurons. *J. Neurosci. Meth.* **93,** 37–48.
8. Estus, S., Zaks, W. J., Freeman, R. S., et al. (1994) Altered gene expression in neurons during programmed cell death: identification of c-jun as necessary for neuronal apoptosis. *J. Cell Biol.* **127,** 1717–1727.
9. Zhou, H., Li, X. M., Meinkoth, J., and Pittman, R. N. (2000) Akt regulates cell survival and apoptosis at a postmitochondrial level. *J. Cell Biol.* **151,** 483–494.
10. Deshmukh, M. and Johnson, E. M. Jr. (1998) Evidence of a novel event during neuronal death: development of competence-to-die in response to cytoplasmic cytochrome c. *Neuron* **21,** 695–705.
11. Putcha, G. V., Deshmukh, M., and Johnson, E. M. Jr. (1999) BAX translocation is a critical event in neuronal apoptosis: regulation by neuroprotectants, BCL-2, and caspases. *J. Neurosci.* **19,** 7476–7485.
12. Neame, S. J., Rubin, L. L., and Philpott, K. L. (1998) Blocking cytochrome c activity within intact neurons inhibits apoptosis. *J. Cell Biol.* **142,** 1583–1593.

13. Loeffler, M., Daugas, E., Susin, S. A., et al. (2001) Dominant cell death induction by extramitochondrially targeted apoptosis-inducing factor. *FASEB J.* **15,** 758–767.
14. Zamzami, N., El Hamel, C., Maisse, C., et al. (2000) Bid acts on the permeability transition pore complex to induce apoptosis. *Oncogene* **19,** 6342–6350.
15. Dokka, S. and Rojanasakul, Y. (2000) Novel non-endocytic delivery of antisense oligonucleotides. *Adv. Drug Deliv. Rev.* **44,** 35–49.
16. Whitt, M., Buonocore, L., and Rose, J. (1987) Introduction of DNA into mammalian cells, in *Current Protocols in Molecular Biology* (Ausubel, F. M., Brent, R., Kingston, R. E., et al., eds.), Wiley Interscience, New York, NY, vol. 1, pp. 9.4.1–9.4.4.
17. Colosimo, A., Goncz, K. K., Holmes, A. R., et al. (2000) Transfer and expression of foreign genes in mammalian cells. *Biotechniques* **29,** 314–331.
18. Leonetti, J. P., Mechti, N., Degols, G., Gagnor, C., and Lebleu, B. (1991) Intracelluar distribution of microinjected oligonucleotides. *Proc. Natl. Acad. Sci. USA* **88,** 2702–2706.
19. Self, A. J., Caron, E., Paterson, H. F., and Hall, A. (2001) Analysis of R-Ras signalling pathways. *J. Cell Sci.* **114,** 1357–1366.
20. LeBlanc, A. C. (1995) Increased production of 4 kDa amyloid β peptide in serum deprived human primary neuron cultures: possible involvement of apoptosis. *J. Neurosci.* **15,** 7837–7846.
21. Terns, M. P. and Goldfarb, D. S. (1998) Nuclear transport of RNAs in microinjected Xenopus oocytes. *Meth. Cell Biol.* **53,** 559–589.
22. Malter, H. E. (1992) Tools and techniques for embryological micromanipulation, in *Micromanipulation of Human Gametes and Embryos* (Cohen, J., Malter, H. E., Talansky, B. E., and Grifo, J., eds.), Raven Press, New York, pp. 250–297.
23. Nicholson, D. W. (2000) From bench to clinic with apoptosis-based therapeutic agents. *Nature* **407,** 810–816.
24. Yuan, J. and Yankner, B. A. (2000) Apoptosis in the nervous system. *Nature* **407,** 802–809.
25. LeBlanc, A., Liu, H., Goodyer, C., Bergeron, C., and Hammond, J. (1999) Caspase-6 role in apoptosis of human neurons, amyloidogenesis, and Alzheimer's disease. *J. Biol. Chem.* **274,** 23,426–23,436.
25a. Stennick, H. R. and Salvesen, G. S. (1997) Biochemical characteristics of caspase-3, -6, -7, and -8. *J. Biol. Chem.* **272,** 25,719–25,723.
26. Harlow, E. and Lane, D. (1999) Staining cells, in *Using Antibodies, A Laboratory Manual* (Harlow, E. and Lane, D., eds.), Cold Spring Harbor Laboratory, Cold Spring Harbor, NY, vol. 1, p. 467.
27. Gervais, F. G., Xu, D., Robertson, G. S., et al. (1999) Involvement of caspases in proteolytic cleavage of Alzheimer's amyloid-beta precursor protein and amyloidogenic A beta peptide formation. *Cell* **97,** 395–406.
28. Pellegrini, L., Passer, B. J., Tabaton, M., Ganjei, J. K., and D'Adamio, L. (1999) Alternative, non-secretase processing of Alzheimer's beta-amyloid precursor protein during apoptosis by caspase-6 and -8. *J. Biol. Chem.* **274,** 21,011–21,016.
29. Lu, D. C., Rabizadeh, S., Chandra, S., et al. (2000) A second cytotoxic proteolytic peptide derived from amyloid beta-protein precursor. *Nature Med.* **6,** 397–404.
30. Weidemann, A., Paliga, K., D rrwang, U., et al. (1999) Proteolytic processing of the Alzheimer's disease amyloid precursor protein within its cytoplasmic domain by caspase-like proteases. *J. Biol. Chem.* **274,** 5823–5829.

Designing
Microarray Experiments
for Neurobiology

P. Scott Eastman and Jeanne F. Loring

1. Introduction

Microarray technology, while it is no longer in its infancy, is still immature and unpredictable. If it works well, microarray expression analysis is a powerful method for estimating the behavior of 10,000 genes simultaneously. But how often does it work well? Because of the rapid adoption of this young technology, we have to anticipate that one of the consequences will be the occasional publication of incorrect results. Some microarray data are simply wrong. In the course of trying to perfect microarray gene expression experiments in the last few years, we think that we have encountered most of the problems that can lead to erroneous interpretation of microarray data. The goal of this chapter is to offer advice from our experience in the hope that it will increase the odds of obtaining accurate data from microarray experiments in neurobiology.

This chapter covers all aspects of the technology, from microarray design to interpretation of data, but it is not intended to be an exhaustive description of all possible methods. Since our experience is primarily with cDNA microarrays, the focus is on techniques that are relevant to that approach. Since our interest is neurobiology, we provide some specific examples of microarray results from cultured neurons and from human and mouse brain tissues. Details of methods that are discussed are provided at the end of the chapter.

From: *Neuromethods, Vol. 37: Apoptosis Techniques and Protocols, 2nd Ed.*
Edited by: A. C. LeBlanc @ Humana Press, Inc., Totowa, NJ

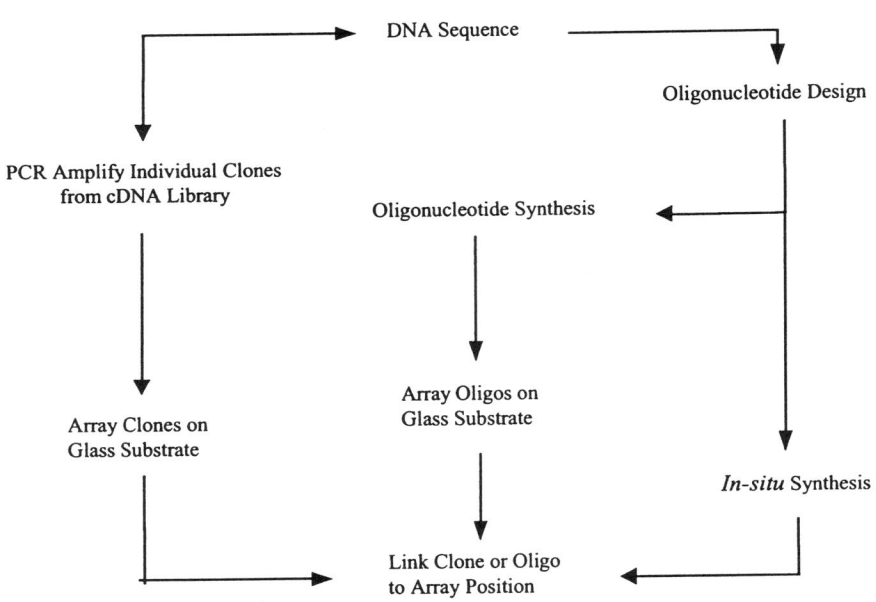

Fig. 1. Flowchart of microarray fabrication.

2. Design of Microarrays

Microarrays are currently available from three sources: commercial suppliers, core facilities, and individual laboratories. Whether or not you have a direct involvement in making the microarray, it is a useful exercise to consider how the array is designed, and what is the impact of the design on the outcome of your experiments. A general overview of microarray fabrication is shown in Fig. 1. Most facilities make microarrays with double-stranded DNA attached to glass. The DNA can be a polymerase chain reaction (PCR) product from a cDNA clone *(1)*, or a PCR product that is directly amplified from genomic DNA by gene-specific primers *(2)*. Another approach uses oligonucleotides that are either synthesized on the substrate or spotted onto glass like cDNAs. The genes can be represented by multiple short oligonucleotides *(3)* or by longer oligonucleotides *(4)*. Ideally, the choice of design of the microarray should be governed by the design of the proposed experiments. The implications of using different types of microarrays are discussed below.

2.1. What Sequences Should Be on the Microarray?

The most fundamental design issue is the selection of the target sequences, which for the purposes of this discussion are defined as the sequences that are attached to a substrate to capture labeled transcripts. Choosing the genes that interest you is the simplest part of the process. It is important to keep in mind that while the current standard is about 10,000 elements per microarray, some of the elements should be internal controls, and if possible, each gene should be represented by at least two sequences. Once you have a list of genes, the representative target sequences must be chosen carefully in order to obtain the desired results. For instance, does the experiment require discrimination of different members of a gene family? If this is a concern, it may be advisable to use target sequences from the 5' or 3' untranslated regions of the mRNA in order to exploit the differences among gene family members. Conversely, if it is more important to do cross-species hybridizations, selecting coding regions of high homology would be better than using the untranslated regions of the message.

Considerations of the abundance of the message and how the mRNA transcripts will be converted into labeled probe also have implications in the selection of target sequences for an array. If the messages of interest are likely to be of low abundance, due either to levels of transcription or to limited sample size or both, amplification of the mRNA population may be necessary. The most commonly used amplification method uses an oligo dT primer that is coupled to a T7 RNA polymerase promoter is annealed to the poly-A sequences of the message. Following first- and second-strand cDNA synthesis, T7 transcription is then employed to amplify the mRNA population (5). The amplified population is by nature 3'-biased. Therefore, the use of 3'-biased sequences in the targets on the glass would be expected to produce optimal results from a T7 RNA polymerase amplified mRNA population. However, it is important to keep in mind that the 3'-untranslated region of some genes contains repetitive sequences, which can lead to undesirable crosshybridization.

Some discussion of sequence verification of target clones is warranted. When considering arrays containing 10,000 or more unique sequences, the possibility of errors in sample handling is enormous. For cDNA microarrays, PCR product is usually generated from cDNA clones that are distributed in a 96-well format, which means that simple errors in handling of the bacterial clones or the purified plasmids can cause crosscontamination of wells. Even if the clones are correctly identified in the wells, it is still possible to select the incorrect plate, or to rotate a plate. PCR of cDNA collections is usually done using vector-specific sequences as universal amplification primers. The chances of error are multiplied by the fact that the same primers will amplify any of the clones. These are not just theoretical issues; it was recently reported that only 62% of a sample of 1189 cDNAs from a supplier of Unigene clones were pure samples of the correct clone (6). Verification of the identity of each clone is clearly necessary, yet resequencing the PCR fragments is beyond the resources of almost any laboratory. Using gene-specific primers to amplify genomic DNA would go a long way toward solving the problem. Another solution to the verification problem is to hedge your bets by spotting at least two clones from each gene cluster. If the multiple clones behave similarly in hybridization experiments, then there is confidence that they are all from the same gene. If they differ, then one of the clones may be incorrect or splice alternatives may be detected. The use of robots, barcodes, and unidirectional plates throughout the process can help to minimize errors in handling. It is important, however, to incorporate quality control measures throughout the process as well as final release criteria to minimize production errors.

2.2. Oligonucleotides, cDNA, or PCR from Genomic DNA?

Although there are currently three different types of DNA sequence that can be used for microarrays, cDNA-based microarrays remain the most cost-effective and flexible format for most core facilities and many commercial suppliers. But using oligonucleotides as target sequence is an alternative to PCR-amplified products. The oligonucleotides can either be synthesized in solution and spotted like cDNA sequences, or synthesized *in situ*, directly on the substrate. Like cDNA microarrays, there are both positive and negative aspects to these methods. The three basic approaches are illustrated in Fig. 2. The cDNA array requires that there be physical clones available, but is relatively simple in concept: the sequences were originally derived from mRNAs, so they should complement them well. Choosing coding regions from genomic DNA requires bioinformatics to choose the appropriate sequences, but since the

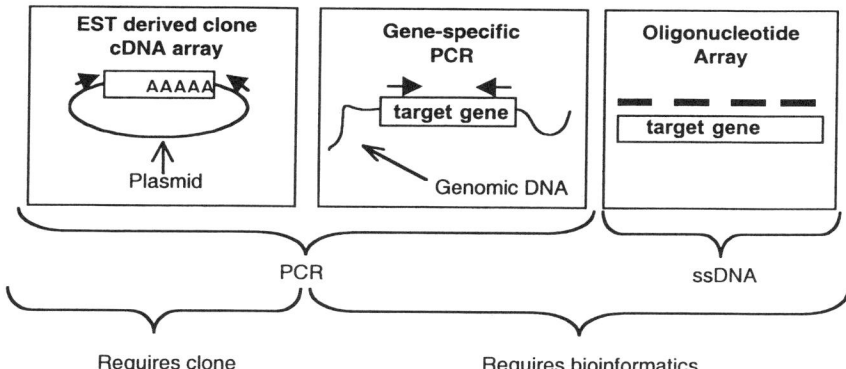

Fig. 2. Types of microarrays.

DNA arrayed will be double-stranded, exon-sized PCR products, the design issues are not complex. Gene specific primers can offer advantages in control of contamination of adjacent wells over vector primers. Oligonucleotide arrays, in contrast, need to be designed more carefully to be sure that the sequences hybridize well with transcripts. There is one fundamental problem unique to oligonucleotide arrays that is not obvious until it actually occurs. Unlike PCR products, oligonucleotides are single-stranded DNA, a feature that offers an additional source of errors. One commercial vendor of *in situ*-synthesized oligonucleotide arrays recently revealed that up to a third of the sequences on a set of mouse arrays were incorrect, reading from the wrong strand of the double helix *(6)*. Performance of an array that uses double-stranded PCR product obviously would not have been affected by this error. Oligonucleotide arrays require bioinformatics and testing to determine which sequences will best represent an mRNA sequence that is several thousands of nucleotides in length. A recent report *(4)* suggested that a good compromise between cDNA clones and short oligonucleotides is 60-nucleotide sequences that represent exons. These long oligonucleotides were synthesized *in situ* by an inkjet printer, which is expensive and available to only a few commercial entities, so this would not be a viable solution for a laboratory or core facility. Unfortunately, the less expensive compromise, to use long oligonucleotides synthesized in a 96-well format, has the same potential for errors in handling as do cDNA arrays.

2.3. Quality Control:
How to Avoid Failures in Manufacturing Microarrays

Building microarrays is an expensive and time-consuming endeavor. But the cost of failure is even more expensive. To go through the entire process of making an array only to find out that a precious sample was wasted on an array that had no hope of working is, to put it mildly, an experience to avoid. It is, therefore, mandatory that there be quality control (QC) measures throughout the process.

The ideal way to be sure that every PCR reaction has worked and has the correct sequence would be to sequence every PCR product. This is far too expensive, so a reasonable compromise is to verify the identity of the clones, then check the PCR products to be sure that they are the right size. Failed PCR amplifications and amplifications with multiple bands can be noted and indicated in the annotations of the array. With subsequent PCR amplifications of these same clones, the size of the amplified product would be required to be within specified QC parameters. Again, those PCR reactions that do not pass the QC criteria are annotated.

It is easy to overlook the effect of having too little or too much target DNA on the array. Insufficient amount of DNA results in suboptimal performance of the microarray, which is not recognized until the hybridization phase *(7)*. Then it is necessary to repeat the PCR amplification, the purification, and the arraying process, which results in doubling of the cost of the array. It should be noted that too much DNA can also have a detrimental effect on array performance, if the amount exceeds the binding capacity of the glass substrate. In this situation, the excess DNA may smear onto the surrounding glass, resulting in elevated backgrounds. Alternatively, the unbound DNA may resolubilize during the hybridization and hybridize to probe in solution, resulting in reduced signal. Consequently, before proceeding to arraying the PCR product, it is prudent to determine the amount of DNA in the samples to be arrayed. Quantification of DNA by optical density at 260 nm is not sufficient because it is too strongly affected by the presence of unincorporated nucleotides and single-stranded DNA. Therefore, we have adopted a method for fluorescent quantification of double-stranded PCR product that is less affected by the presence of contaminating species. We recommend a picogreen fluorescent assay (Molecular Probes, Eugene, OR) for measuring the amount of DNA before arraying.

2.4. Making Sure the Array Works: Controls, Controls, Controls

It may be a surprise that the first QC measure after arraying may not be a hybridization. This is because if you hybridize with probe generated from a complex RNA, there will most likely be insufficient signal for each element to allow QC of all of the elements. However, if there is insufficient signal in a QC hybridization with a complex probe, how can you tell whether the problem lies with the array or with the probe? To test the array a non-hybridization-based measure of the amount of DNA retained on the glass following arraying and postprocessing will provide the best measure of the transfer of DNA from the wells to the glass substrate. We have adapted a fluorescent DNA assay from cell culture techniques to quantify the amount of DNA retained on the array. This method uses Syto (Molecular Probes, Eugene, OR) staining of the array to allow quantification of the amount of DNA following scanning in the Cy5 channel.

Another useful measure for qualification of an array is possible if the PCR products are made by amplification with primers to common vector sequences in the clones. Hybridization of the array with an oligonucleotide specific for a common vector sequence labeled with one of the Cy dyes can provide a measure of the amount of DNA available for hybridization in each element. Only after these tests should a probe generated from a complex RNA population be used to confirm that the target DNA is in a state that it is possible to hybridize.

3. Preparing Samples

Once an array is produced and qualified, the experimental samples are prepared, labeled, hybridized, and the data acquired. The steps through these processes are outlined in Fig. 3. We will focus only on methods that are relevant to cDNA microarrays using competitive hybridization.

3.1. Isolating RNA

A useful guideline for competitive hybridizations is to isolate at least 200 ng of mRNA from each sample. Commercially available kits for the isolation of total RNA and poly-A selection of messenger RNA work well. The isolation of RNA from cultured cells is a straightforward process and usually 10^6–10^7 cultured cells

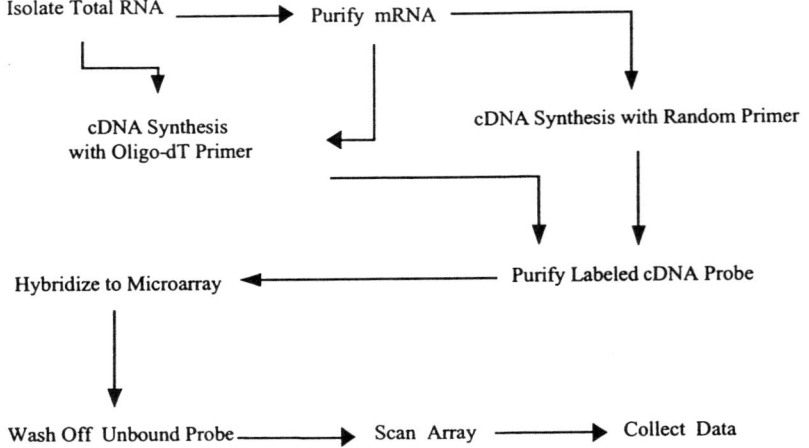

Fig. 3. Flowchart of cDNA microarray experiment.

provide sufficient mRNA for a set of microarray experiments. The isolation of RNA from tissue samples, however, has some additional considerations. It is usually possible to isolate mRNA from human brain samples that are frozen within 24 h postmortem *(8)*. Mouse brain samples can be isolated under controlled conditions and yield predictable amounts of mRNA. The yields of mRNA isolated from mouse tissue using the methods described later in this chapter are presented in Table 1.

A variable that is easy to overlook is the precision of dissection of the tissue. The presence of contaminating tissue can sometimes be seen in the microarray results. For example, the exceptionally high level of transthyretin mRNA that appears occasionally in samples of mouse cortex is probably due to inclusion of choroid plexus tissue; choroid plexus is the major site of transthyretin synthesis in the brain. Microarray results comparing gene expression in cortex from two mouse brains are shown in Fig. 4. One of the samples expresses far higher transthyretin levels than the other, indicating that one of the samples probably contained contaminating choroid plexus. There are a number of ways to increase accuracy of dissection. The most low-tech method is to cut thick sections of the tissue and then dissect along anatomic boundaries. For smaller regions, laser capture microdissection can be used to isolate small numbers of cells from thin slices of frozen tissue. The small samples will have to be amplified using a method such as T7 amplification.

Table 1
RNA Yields from Mouse Tissues

Mouse tissue	Tissue weight (mg)	Total RNA (µg)	1st selection polyA RNA (µg)	2nd selection polyA RNA (µg)
Cerebellum 1	320	250	10.5	2.3
Cerebellum 2	330	208	8.5	3.0
Cerebellum 3	310	230	10.0	3.0
Cerebellum 4	320	163	7.0	2.4
Spleen 1	280	425	8.8	2.6
Spleen 2	370	525	7.2	2.1
Spleen 3	230	325	8.8	2.1
Spleen 4	280	400	9.6	3.1

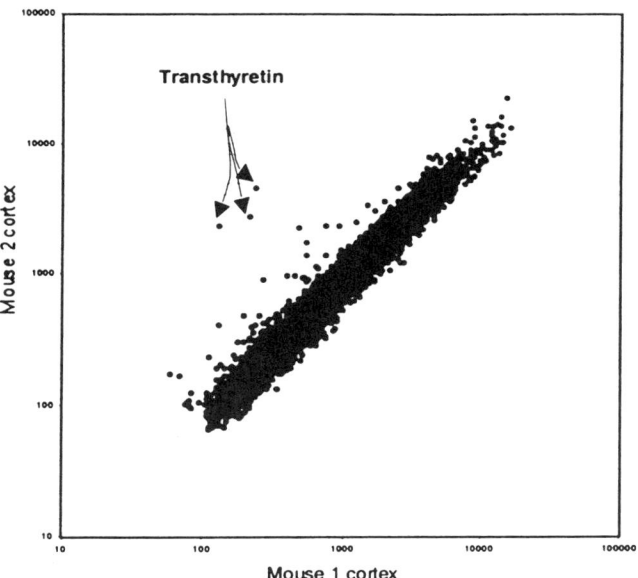

Fig. 4. Gene expression resulting from dissection of contaminating tissue.

3.2. Quality Control:
How to Avoid Failures in Probe Labeling

One of the most problematic sources of failed hybridizations is the quality and quantity of the RNA used in the labeling reaction. As in the other procedures, stringent QC measures introduced in the labeling step greatly decrease the number of hybridization failures.

The random labeling method (*see* Subheading 6.) requires iso-lation of mRNA and elimination of as much ribosomal and trans-fer RNA sequences as possible. To isolate mRNA, we routinely use a commercially available kit for selection of poly-A messen-ger RNA (Qiagen, Valencia, CA). A second round of poly-A selec-tion only occasionally improves the purity of the mRNA preparation, and results in reduced yields of mRNA (*see* Table 1). As a first step following poly-A selection, an electrophoretic profile of the RNA is obtained to assess the amount of rRNA contamination and any degradation of the sample. Standard agarose gel electrophoresis is sufficient to assess the quality of mRNA; for rapid quantitative mRNA profiles, we use a microcapillary electrophoresis system (Agilent 2100 Bioanalyzer; Agilent, Palo Alto, CA). Inadequate removal of rRNA has a detrimental effect on the quality of the data from a hybridization. The impact of titrating total RNA into poly-A selected RNA can be seen in Fig. 5. With large amounts of con-taminating RNA in one channel, many elements can appear to be differentially expressed in a hybridization between identical mRNA populations in which there should be no differential expression.

Qualified poly-A RNA must also be accurately quantified before using it for hybridization. Due to the inherent problems with unin-corporated nucleotides resulting in elevated OD260 nm readings, we use a fluorescent Ribogreen (Molecular Probes, Eugene, OR) method of quantifying RNA. Increasing amounts of RNA input in a labeling reaction results in increasing amounts of signal up to a plateau. The impact of differing amounts of RNA put into a labeling reaction can be seen in Fig. 6. As one would expect, differ-ent amounts of RNA input into the labeling reactions results in differing amounts of signal in the two channels, which causes problems with interpretation of hybridization data (see below).

3.3. Labeling mRNA for Detection

Messenger RNA is commonly labeled by primer extension either from poly-A sequences or by random priming and incorporation of labeled nucleotides or the use of labeled primers. We have cho-sen to use a labeling system consisting of cDNA synthesis with labeled random oligonucleotides. This has provided us with more signal than poly-A priming with labeled nucleotide incorpora-tion, and offers flexibility to use either 3'- or 5'-biased target cDNA sequences. Also, as indicated above, if oligo dT primers are used, the sequences will be 3'-biased, and that should be taken into con-

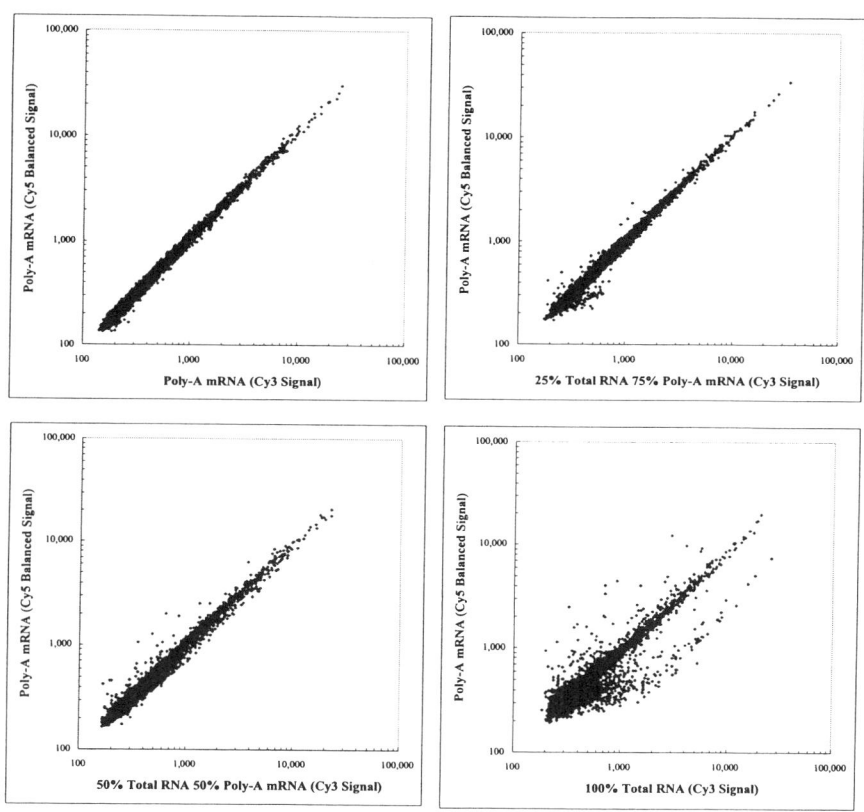

Fig. 5. Effect of contamination of mRNA with total RNA.

sideration when designing the array and interpreting hybridization results.

The effects of differences in labeling efficiency between Cy3 and Cy5 primers are similar to the impact of differences in RNA input (see above). To avoid this problem, the Cy3 and Cy5 primers need to be tested to determine whether the primers label RNA with the same efficiency.

3.4. Hybridization: The More Controls, the Better

Most cDNA microarray protocols are based on competitive hybridization of two samples of mRNA: a control and an experimental sample, each labeled with a different fluorescent tag. We prefer this approach to single-sample hybridizations for several

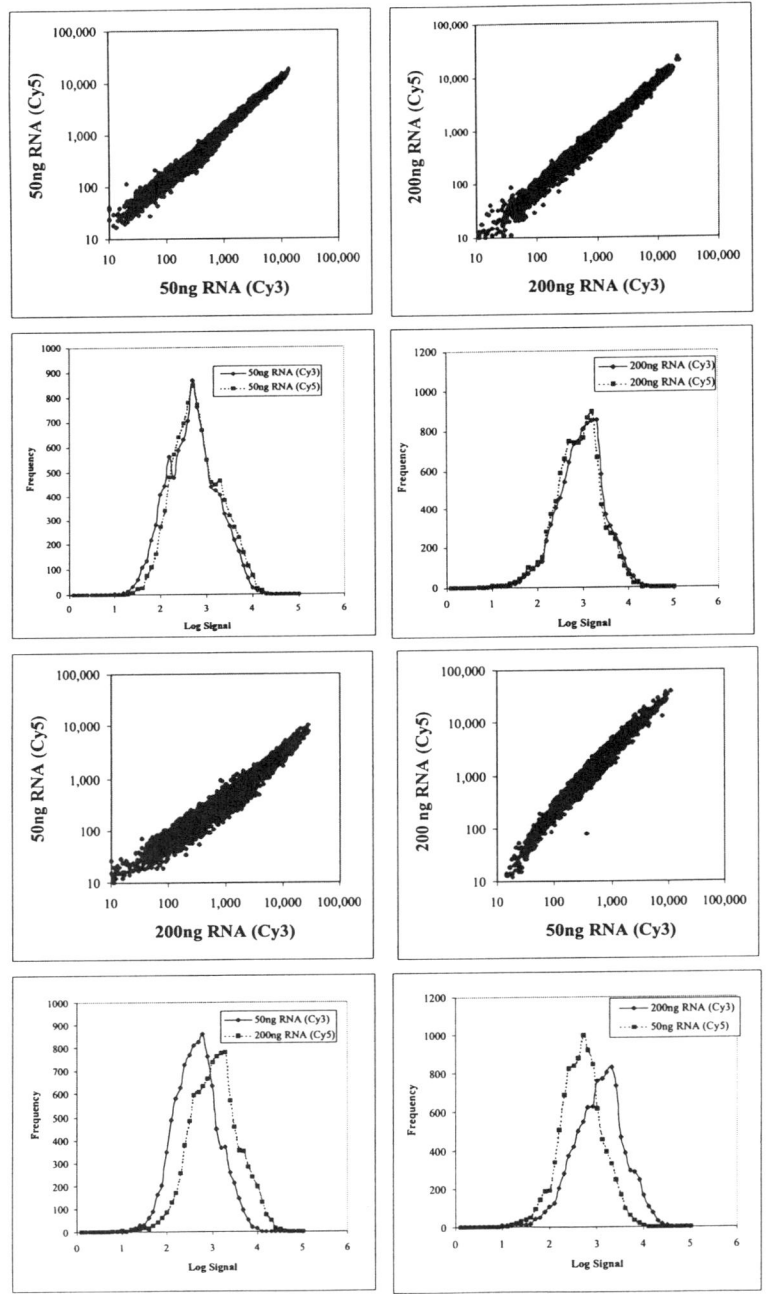

Fig. 6. Effect of quantitative imbalance in competitive hybridization.

reasons. For example, by directly controlling for variables, it fits well into classic experimental designs. Also, the significance of results is much improved by repeating an experiment by "flipping the channels," which is discussed below. When creating a database of a large number of gene-expression experiments, the competitive hybridization approach compensates for variations in the microarrays and the sample preparation. When the two samples "match," as explained below, and controls are included for normalization, the hybridizations are very reproducible.

We recently reported *(7)* that one of the greatest sources of variation in microarray experiments was the hybridization itself. In order to estimate the linearity, sensitivity, and reproducibility of the hybridization, onboard controls were used to serve as a QC assay for each array. Controls are included in each labeling reaction and each hybridization. Control elements to be arrayed can be obtained by PCR amplification of nontranscribed regions of the genome from a species evolutionarily distant from the one to be studied, for example, bacterial controls for human arrays. For positive controls, RNA transcripts corresponding to the array control elements can be generated by appending a T7 RNA polymerase promoter onto one of the PCR primers and performing a T7 transcription reaction. Following qualification and quantification of several of these transcripts, they can be spiked into the labeling reaction at various levels to monitor the sensitivity and the dynamic range of the system on each array. Additionally, such T7 transcripts can be designed to include a poly-A sequence that can be used to monitor RNA extraction and/or mRNA preparation.

When designing controls for a two-channel system, it is important to determine the linearity, sensitivity, and dynamic range of the signal in each channel independently. However, it is also necessary to control for the quality of differential expression that is measured using both channels. A simple way to do this is to spike T7 RNA transcripts into the labeling reaction in differing amounts in the two channels to ask whether differential expression ratios are faithfully reported at different signal levels.

QC is useful for more than determining the linearity and sensitivity of the hybridization. It is a powerful tool to identify the source of problems when a hybridization goes awry. For example, absence of signal in one channel can have many causes. Absence of signal in both the onboard controls and the sample RNA may indicate a problem with the labeling reaction or contamination of the sample RNA with an inhibitor or RNase. Absence of signal in the sample RNA but not in the onboard controls would indicate proper functioning of the labeling reaction, but a problem with the sample RNA.

The most important lesson we have learned about competitive hybridizations is that the control and experimental samples must be balanced. If there are equal amounts of labeled transcript in each channel, hybridizations using the same mRNA for both channels (a self/self hybridization) are consistent from 50 ng to 400 ng input. An imbalance in signal leads to several problems with the results. The impact of unequal amounts of RNA in the two channels of a self/self hybridization is shown in Fig. 6. Imbalance causes an increase in the variability of differential expression, which limits the ability to discriminate differential expression. Also, imbalance causes deviation of the data from the 45° angle expected for a plot of Cy3 vs. Cy5 signals, which especially affects the ability to compare hybridizations such as in a time course analysis. An imbalance can be detected by comparing histograms of signal intensity in each channel across a hybridized microarray. It is more important to have matching inputs than to have large amounts of sample. For example, if only 50 ng of one sample is available, it should be hybridized with 50 ng of the other sample, even though 200 ng is the standard amount used and an unlimited amount of the second sample is available.

3.5. Detecting the Hybridization to a Microarray

Scanners for detection of hybridization in both the Cy3 and Cy5 channels are commercially available; scanner performance has been evaluated in depth in published reviews *(9)*. We currently use a scanner produced by Axon Instruments (Foster City, CA). On most scanners, the only adjustment that can be made by the user is the photomultiplier tube (PMT) voltage setting. There is a temptation to change the PMT voltage for individual scans in order to balance the signal from the two channels. This should only be necessary if the two samples were not balanced by mRNA input and label incorporation. Changing PMT settings may change the signal to noise ratio of the output and increasing a PMT setting to increase signal results in a corresponding increase in background. High background levels reduce the ability to discriminate differences in signals. Optimal PMT settings, which balance the two channels, are obtained with equimolar amounts of the Cy3 and Cy5 dyes providing comparable signal throughout the signal spectrum. In addition, it is important to monitor the performance of the scanner. Note that the lasers employed to generate the fluorescence will lose power over time.

3.6. Interpretation of Microarray Data: What's Real?

Microarray data analysis is population-based, and replicates and statistics are required to extract the reliable information. The reproducibility of cDNA microarray hybridizations was tested in a recent report (7) in which mRNA was converted to labeled cDNA probe in 10 separate pairs of reactions and competitively hybridized to identical cDNA microarrays. The results of this analysis showed that no individual array element gives the same signal or differential expression value in all 10 of 10 hybridizations. Instead, the 10 data points for each element are a population of values that can be described by a mean and standard deviation. This means that since the values are not exactly the same each time, comparisons of experimental conditions can only be made in the context of a statistical analysis of replicated data for each time point.

There has been much discussion of the impact of false positives on the ability to identify "truly" differentially expressed genes (10). We found in the replicate study described above that 99% of the elements on each array fell within the 99.5% tolerance interval, which was plus or minus 1.4-fold differential expression (7). Put another way, this means that for a 10,000-element array, 100 elements would be expected to fall outside the tolerance interval and thus be falsely identified as differentially expressed. We have observed that the greatest variance in microarray data comes not from an obvious variable such as the level of expression of a gene, but from the unique behavior of each element. This observation is illustrated in Fig. 7, where the range of differential expression across 10 replicate determinations is quite large for some genes and some genes have a narrow distribution of differential expression regardless of their level of differential expression.

An additional consideration with the two-color system is the possibility of bias in one channel (7), such that the same transcript labeled with one of the dyes hybridizes differently when labeled with the other dye. Luckily, this is rare and minimal. There is a simple way to control for bias in a dual-channel system: a reflective labeling experiment, or channel flip, in which each sample is divided and labeled separately with each of the dyes. While eliminating

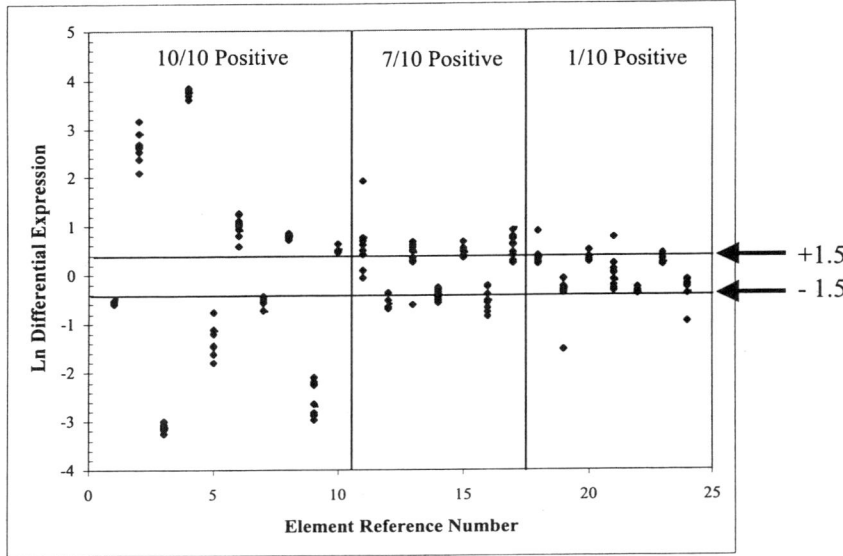

Fig. 7. Distribution of differential expression values in 10 replicate microarrays.

any errors of bias, this approach has the added benefit of increasing the confidence level by serving as a replicate hybridization. It should be noted that there is no straightforward way to judge the bias in a single-channel system.

3.7. How Many Replicates Do I Need?

It is clear from the observations discussed above that microarray experiments *must* be replicated. But how many replicates are necessary? From a statistical point of view, the more replicates, the better the data. Pragmatically, however, the amount of sample and the cost of the arrays may be limiting.

The number of replicates necessary depends on the system used, the overall variability of the system, the variability of the individual genes, and, most important, the design of the experiment. We have taken an empirical approach based on data from a group of 10 heterotypic hybridizations (7). We used the same preparation of mRNA, but maximized the other sources of variation by labeling separately for every hybridization and using different printing batches of microarrays. In a perfect world, one would expect, regardless of the cutoff employed to define differentially expressed

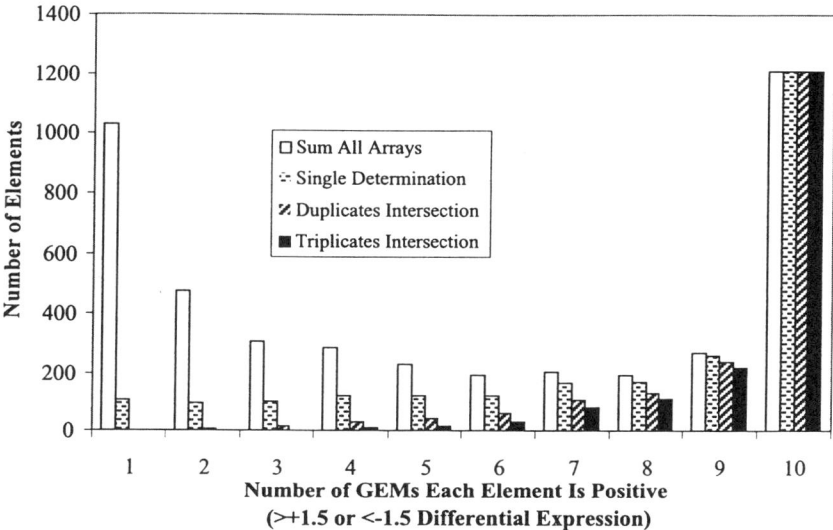

Fig. 8. Distribution of element differential expression: effect of duplicate and triplicate measurement.

elements, that all of the elements would either be differentially expressed in all 10 hybridizations or in none of the 10 hybridizations. In reality, however, there will be some number of elements that are identified as differentially expressed in only one of the 10 hybridizations (*see* Fig. 7).

In Fig. 8, the number of times that an element shows differential expression is plotted against the number of elements that fall into that group. The data are distributed along an inverse bell curve, and as a population are independent of the defined cutoff within a reasonable range. Of course, if the cutoff were arbitrarily set at 10,000-fold, all of the elements would be in the 0-out-of-10 bin, and if it were set at 1-fold, all would be in the 10-out-of-10 group.

The important message from the replication experiment is that if a single hybridization were performed, there would be some apparently regulated genes that would fail to show differential expression again even if the experiment were repeated nine more times. Follow-up evaluation of such genes would be costly and most likely fruitless.

Performing a single replicate, preferably a biological replicate performed with reflective labeling, and requiring that the replicates both be identified as differentially expressed, will eliminate all the elements that were positive on only one hybridization and greatly reduce the number of elements that would be replicated only five times in a 10-array experiment. The price of requiring the same result on both arrays will be a small loss of elements that would have been replicated nine, eight, seven, or six times if a 10-array experiment had been performed. Should more than two replicates be done? Using the data from the 10-replicate experiment, we determined that there is only a marginal improvement in the data if the number of replicates is increased from two to three (Fig. 8).

The take-home message from our results is that more is better, but if the caveats discussed above are taken into consideration, duplicate experiments should be sufficient to address most biological questions. Repeating the experiment three times or more increases the reliability of the data, but since each extra replicate provides only a small increment of improvement in the results, it might be wiser to use the additional arrays to measure another biological variable.

3.8. What Can Be Done to Deal With a Very Large Amount of Data from Multiple Experiments?

The analysis of a complete experimental design dataset, for example, a study comparing experiments and controls over several time points, most often employs the use of cluster analysis. Clustering of data allows the discrimination of patterns that may not have been obvious in unanalyzed data. It may, for example, show that three genes demonstrate similar behavior under a wide variety of conditions, which suggests that the genes may be functionally linked. We will not recommend a particular product, but there are many commercial analysis packages that differ mostly in the user interface. The most commonly employed methods of clustering are the *K* means and hierarchical clustering. The advantage of the hierarchical clustering is that, as opposed to *K* means, there is no need to *a priori* designate the number of clusters. The disadvantage of the hierarchical clustering is that it is calculation-intensive; consequently, it may not be possible to cluster many genes at a time and the runtime may be long. While hierarchical clustering provides several different potential clusters, there is currently no statistical method for determining the optimal number of clusters. Therefore it may be necessary to test different numbers of clusters and to make a judgment of what appears to the most informative clustering.

In some situations, such as the construction of a large expression database, or if samples are limited, it may not be possible to perform all of the necessary comparisons by direct competitive hybridization. In this case, different hybridization experiments may need to be compared through "virtual hybridization." This is a way of using the signals from single channels in a competitive hybridization to compare indirectly with signals from another hybridization. In order to make such a comparison, the data from all of the hybridizations have to be normalized so that signal dynamic ranges from different data sets are matched (see above). Several methods for normalization have been developed. For example, a ranking by comparison of the signal value of each gene to the mean of the population is helpful but is limited in that it does not take into account the dynamic range of the population. More commonly employed is a Z score, which is the residual or difference between the individual value and the mean compared to the standard deviation of the population performed in log space. One of the implicit assumptions of using these methods is that the normalized data has a near-normal distribution, which may not always be true *(10)*.

We tested a "virtual hybridization" approach using the set of 10 competitive hybridization experiments discussed in an earlier section. Ten control hybridizations using samples from the same mRNA population showed that at the 99.5% tolerance interval, 99% of the elements fell within a 1.4-fold cutoff for differential expression. As discussed above, this does not mean that a difference of greater than 1.4-fold is necessarily a real difference, but instead means that elements that show less than a 1.4-fold change cannot be identified as being differentially expressed. A similar analysis should be performed on a "virtual" hybridization. To do this, we performed "virtual" self/self hybridizations of placental mRNA by comparing the signals from 20 competitive hybridizations of placental RNA with RNA from heart. A histogram of the differential expression of the actual hybridizations and the virtual hybridizations is shown in Fig. 9. Those data indicated that the minimal detectable differential expression for the virtual hybridizations is 2-fold rather than 1.4-fold. This means that if the experiments are replicated sufficiently to do "control virtual" hybridizations, the set of genes identified as being differentially expressed will be a subset of those identified through actual hybridizations.

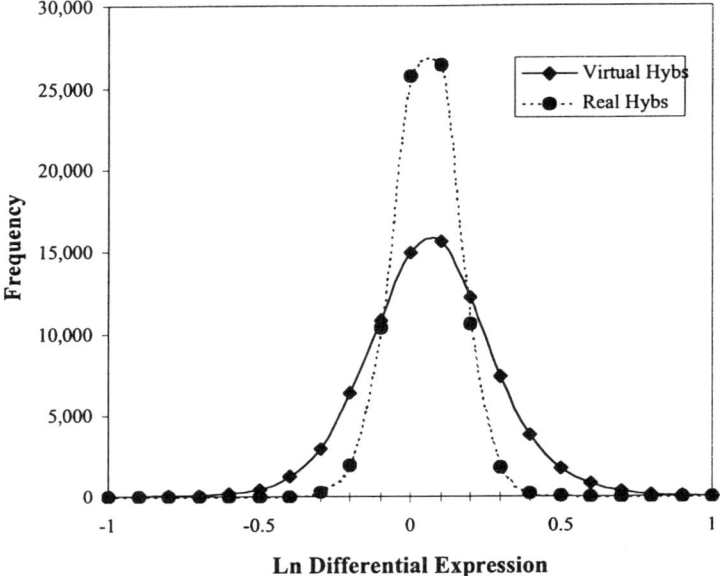

Fig. 9. Comparison of virtual and actual competitive hybridizations.

3.9. Do the Results Need
to Be Confirmed by a Different Method?

We have argued that in their current state of development, micro-arrays are best used as a screening tool to identify genes that show interesting behavior in experiments. Eventually all microarray technology will be sufficiently accurate so that the results will be accepted as definitive. Until that is accomplished, however, it is sometimes wise to confirm microarray results by a different, quantitative technique. The most common methods of confirming microarray data are Northern blot analysis and real-time PCR analysis. It should be noted that direct comparison of these methods to microarray analysis may not be straightforward. Determination of differentially expressed genes in a microarray experiment involves a comparison with the average signal of a large number of genes. It is not practical to perform the same type of a population-based analysis with either Northern blot or real-time PCR, so instead the signals are compared to "housekeeping" genes that are believed to be invariant in all tissues. We want to add a cautionary note about the use of this approach: an evaluation of hundreds of microarray experiments has failed to identify *any* gene that is invariant in expression across a large number of tissues. Therefore, the housekeeping genes that are chosen as comparison controls for Northern analysis or real-time PCR should be operationally defined as those that are consistently invariant across the tissues that are to be analyzed. These genes could be chosen on the basis of their behavior on a set of microarrays performed on the biological system of interest.

4. Examples of Microarray Results in Neurobiology

4.1. Mouse Embryonic Stem Cells

We used a mouse microarray (Incyte's Mouse LifeArray) to ask what changes occur during retinoic acid-induced differentiation of ES cells *(11)*. We compared the line of ES cells (E14) with a line derived from a subclone in the course of a gene targeting experiment. As can be seen in Fig. 10A, the two undifferentiated cell lines were identical with respect to gene expression pattern. However, after the cells were induced to differentiate, primarily into neurons, the gene expression pattern changed greatly (Fig. 10B). We detected consistent differential expression of 92 of the sequences in five separate experiments. Most of the genes that were downregulated were involved in protein synthesis and replication. The upregulated transcripts included neuron-specific genes and extracellular matrix components.

4.2. Human Alzheimer's Disease

We used a human microarray (Incyte's Human Unigene LifeArray) to compare gene expression patterns in control and moderate to severe Alzheimer's disease (AD) brains *(12)*. Postmortem intervals were all less than 24 h for the six AD samples and nine controls. The analysis of these data was considerably more complicated than the analysis of data from the cultured ES cells, because of the biological variability among human brain tissues. We dealt with differences between individuals by comparing pairs of controls, and removing from further analysis all of the genes that were differentially expressed between controls. An example of a control–control hybridization is shown in Fig. 10C. This stringent filtering gave us confidence that we were excluding gene expression differences that were not associated with AD. We compared age-matched

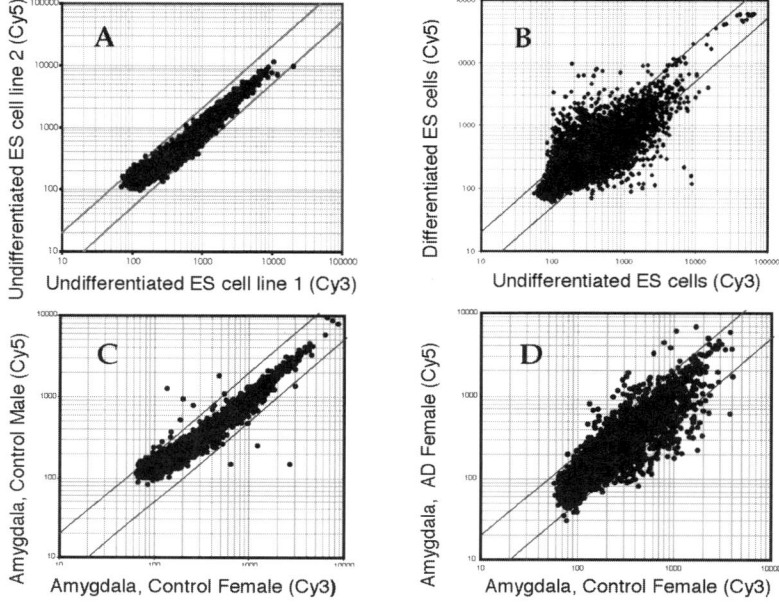

Fig. 10. Scatter plots of data from microarray experiments.

pairs of AD and control cingulate cortex, amygdala, striatum, and cerebellum by competitive hybridization. The genes that are most likely to be associated with AD pathology were those that were differentially expressed only in the amygdala and/or cingulate cortex. The final list of candidates numbered 118 genes. An example of a comparison between control and AD amygdala is shown in Fig. 10D.

5. Conclusions

By the time this chapter is published, there will probably be another 100 publications of data using microarrays. If each paper focuses in on the 100 most interesting genes that are differentially expressed, that means that 100,000 reports of differential expression will have been made. If our estimates hold true, then 1 percent, or 1000 determinations, will simply be wrong.

This should not deter anyone from performing microarray experiments, and as databases of information are built, much of this error should be sorted out by repetition of the experiments. A useful analogy is the GenBank database, in which huge amounts of data are added every day, and weeding out the incorrect information has become the curators' major task.

Just as DNA sequencing has improved because of the pressure of scientists using the technology, so will large-scale gene expression technologies become better, cheaper, and faster. However, the most important thing to keep in mind is that it is the mind of the biologist that is the most important ingredient in the application of any technology. If the experiments are designed well, microarrays will give us the answers.

6. Methods

6.1. Synthesis of PCR Products

PCR was used to generate large quantities of defined target DNA for microarray production *(7)*. Plasmids containing cloned genes were grown in *Escherichia coli* and were amplified using vector primers SK536 and SK865. Briefly, 1 µL of bacterial cell culture was added to reaction buffer containing 10 mM Tris-Cl, pH 8.3, 1.5 mM MgCl$_2$, 50 mM KCl, 0.2 mM each dNTP, 0.5 µM each primer, and 2 units Taq polymerase. The mixture was incubated for 3 min at 95°C, and 30 cycles of PCR were performed at 94°C/30 s, 56° C/30 s, and 72°C/90 s. PCR products were purified by centrifugal chromatography with Sephadex S400 resin (Amersham-Pharmacia Biotech, Uppsala, Sweden) in 0.2X standard saline citrate buffer (SSC) in a 96-well format. Purified PCR products were concentrated to dryness and resuspended in 10 µL H$_2$O.

6.2. Qualification and Quantification of PCR Products

PCR products were analyzed by agarose gel electrophoresis, and samples that failed to amplify or had multiple bands were annotated. PCR products were quantified using PicoGreen® dye (Molecular Probes, Eugene, OR) according to the manufacturer's instructions.

6.3. Arraying and Postprocessing

Ten thousand PCR products were arrayed by high-speed robotics. Each element occupied a spot of approximately 150 µm in diameter and spot centers were 170 µm apart. DNA adhesion to the glass was achieved by irradiation in a Stratalinker model 2400 UV illuminator (Stratagene, San Diego, CA) with light at 254 nm and an energy output of 120,000 µJ/cm^2. The microarrays were washed for 2 min in 0.2% sodium dodecyl sulfate (SDS) followed by 3 rinses in H$_2$O for 1 min each. The microarrays were treated with 0.2% (w/v) I-block (Tropix, Bedford, MA) in phosphate-buffered saline (PBS) for 30 min at 60°C. They were washed again for 2 min in 0.2% SDS, rinsed 3 times in H$_2$O for 1 min each, and finally dried by a brief centrifugation.

6.4. Array Qualification: SYTO® 61 Dye

The SYTO 61 nucleic acid staining procedure was modified to allow its use for measurement of DNA bound to microarrays. A 5-µM stock solution of SYTO 61 dye (Molecular Probes, Eugene, OR) was diluted 1/100 in 10 mM Tris-HCl, pH 7, 0.1 mM EDTA (TE). Microarrays from each manufactured batch were immersed in this solution for 5 min at room temperature, rinsed with TE, rinsed with H$_2$O, and finally with absolute ethanol. The microarrays were scanned on a GenePix 4000A scanner (Axon Instruments, Foster City, CA) at 625 nm.

6.5. mRNA Preparation and Probe Synthesis

Total cellular RNA was prepared with Trizol reagent (Life Technologies, Gaithersburg, MD) from mouse ES cells (11), human brain tissue (12) and mouse brain (Loring in prep). mRNA was isolated from total RNA from these sources and from commercially available placenta, brain, and heart total RNA (Biochain, San Leandro, CA) by a single round of poly-A selection using Oligotex resin (Qiagen, Valencia, CA). The purified mRNA was quantified using RiboGreen dye (Molecular Probes, Eugene, OR). RiboGreen dye was diluted 1/200 (v/v final) and mixed with RNA at concentrations ranging from 1 to 5000 ng/mL as determined by absorbance at 260 nm. Millennium RNA size ladder (Ambion, Austin, TX) was used to generate standard curves. Fluorescence was measured in 96-well plates with a FLUOstar fluorometer (BMG Lab Technologies, Germany) fitted with 485-nm (excitation) and 520-rm (emission) filters.

Twenty-five to 100 ng of mRNA were qualified on an Agilent 2100 Bioanalyzer (Agilent Technologies, Palo Alto, CA), to examine the mRNA size distribution. Purified mRNA was converted to Cy3- or Cy5-labeled cDNA probe using a custom labeling kit (Incyte Genomics, Fremont, CA). Each reaction contained 50 mM

Tris-HCl, pH 8.3, 75 m*M* KCl, 15 m*M* MgCl$_2$, 4 m*M* dithiothreitol (DTT), 2 m*M* dNTPs (0.5 m*M* each), 2 µg Cy3 or Cy5 random 9-mer (Trilink, San Diego, CA), 20 U RNase inhibitor (Ambion, Austin, TX), 200 U MMLV RT RNase (H-) (Promega, Madison, WI), and mRNA. Labeled Cy3 and Cy5 cDNA products were combined and purified with a size-exclusion column, concentrated by ethanol precipitation, and resuspended in hybridization buffer.

6.6. Array Qualification: Complex and Vector-Specific Hybridizations

Hybridization of labeled cDNA probes was performed in 20 µL 5X SSC, 0.1% SDS, and 1 m*M* DTT at 60°C for 6 h as previously described *(7)*. Hybridization with a Cy3-labeled vector-specific oligonucleotide (Operon Technologies) was performed at 10 ng/µL in 5X SSC, 0.1% SDS, 1 m*M* DTT at 60°C for 1 h. The microarrays were washed after hybridization in 1X SSC, 0.1% SDS, 1 m*M* DTT at 45°C for 10 min, and then in 0.1X SSC, 0.2% SDS, 1 m*M* DTT at room temperature for 3 min. Following drying by centrifugation, microarrays were scanned with a Axon GenePix 4000A fluorescence reader (Axon Instruments, Foster City, CA) at 535 nm for Cy3 and at 625 nm for Cy5. An image analysis algorithm in GEMTools™ software (Incyte Genomics) was used to quantify signal and background intensity for each target element. The ratio of the two corrected signal intensities was calculated and used as the differential expression ratio for this specific gene in the two mRNA samples.

The Axon scanner was calibrated using a primary standard and a secondary standard as previously described *(7)*. For the primary standard, hundreds of probe samples were prepared which were fluorescently balanced in Cy3 and Cy5 channels as determined by a Fluorolog3 fluorescence spectrophotometer (Instruments S.A., Edison, NJ). These probes were hybridized to microarrays and the scanner PMTs were adjusted to give balanced fluorescence and the greatest dynamic range. Using these PMT values, a fluorescent plastic slide was scanned to obtain corresponding fluorescent values. This secondary standard was used to calibrate other scanners on a daily basis.

Acknowledgments

We thank Drs. Bruce Wang, Julia Montgomery and Cathleen Chan for contributing figures for this chapter, and for providing us with their unpublished data. We also want to thank Dr. Mark Reynolds for his helpful discussions and Dr. Jeff Seilhamer for his support of this work. We thank the technical staff of the microarray facility of Incyte Genomics for providing the infrastructure that made this work possible. We reserve a special kind of gratitude for Dr. Rick Johnston, who continues to be an inspiration to us in many ways.

References

1. Iyer, V. R., Eisen, M. B., Ross, D. T., et al. (1999) The transcriptional program in the response of human fibroblasts to serum. *Science* **283,** 83–87.
2. Penn, S. G., Rank, D. R., Hanzel, D. K., and Barker, D. L. (2000) Mining the human genome using microarrays of open reading frames. *Nature Genet.* **26,** 315–318.
3. Lockhart, D. J., Dong, H., Byrne, M. C., et al. (1996) Expression monitoring by hybridization to high-density oligonucleotide arrays. *Nature Biotechnol.* **14,** 1675–1680.
4. Hughes, T. R., Mao, M., Jones, A. R., et al. (2001) Expression profiling using microarrays fabricated by an ink-jet oligonucleotide synthesizer. *Nature Biotechnol.* **19,** 342–347.
5. Van Gelder, R. N., von Zastrow, M. E., Yool, A., Dement, W. C., Barchas, J. D., and Eberwine, J. H. (1990) Amplified RNA synthesized from limited quantities of heterogeneous cDNA. *Proc. Natl. Acad. Sci. USA* **87,** 1663–1667.
6. Knight, J. (2001) When the chips are down. *Nature* **410,** 860–861.
7. Yue, H., Eastman, P. S., Wang B. B., et al. (2001) An evaluation of the performance of cDNA microarrays for detecting changes in global mRNA expression. *Nucleic Acids Res.* **29,** e41–e1.
8. Ross, B. M., Knowler, J. T., and McCulloch, J. (1992) On the stability of messenger RNA and ribosomal RNA in the brains of control human subjects and patients with Alzheimer's disease. *J. Neurochem.* **58,** 1810–1819.
9. Worley, J., Bechtol, K., Penn, S., et al. (2000) Systems Approach to Fabricating and Analyzing DNA Microarrays, Eaton, Westborough, MA.
10. Manduchi, E., Grant, G. R., McKenzie, S. E., Overton, G. C., Surrey, S., and Stoeckert, C. J. Jr. (2000) Generation of patterns from gene expression data by assigning confidence to differentially expressed genes. *Bioinformatics* **16,** 685–698.
11. Loring, J. F., Porter, J. G., Seilhamer, J., Kaser, M., and Wesselschmidt, R. (2001) Gene expression profile of embryonic stem cells. Restorative. *Neurol. Neurosci.* (in press).
12. Loring, J. F., Wen, X., Seilhamer, J., Lee, J., and Somogyi, R. (2001) A gene expression profile of Alzheimer's disease. *DNA Cell Biol* (in press).

Analysis of Gene Expression by Nylon Membrane cDNA Arrays

H. Michael Tucker and Steven Estus

I. Introduction

The use of DNA arrays offers the promise of semiquantitative analysis of large numbers of genes simultaneously. Here, we report our efforts at applying this technology to an issue of potentially high relevance to current research in Alzheimer's Disease (AD). The AD field is in strong need of biomarkers suitable for the diagnosis of AD prior to symptomology. Amyloid-beta (Aβ) deposition in the brain is a hallmark of AD, and amino acid racemization studies suggest that Aβ deposition occurs well before symptomology *(1)*. Hence, approaches that provide an indirect indication of Aβ burden may be useful in early AD diagnosis. A mouse model of Aβ deposition resulting from overexpression of the Aβ protein precursor has proven utility as a model of AD. Although such mice do not have the striking neuronal loss seen in AD, these mice manifest many of the other attributes of AD including fibrillar and nonfibrillar Aβ deposits, neuritic plaques, decreased synapse density and gliosis, and behavior problems *(2–4)*. Moreover, they have proven useful in studies aimed at decreasing Aβ deposition, e.g., the Aβ vaccination studies *(5–7)*. As a means to identify markers of Aβ deposition that may prove useful in diagnosing AD before symptomology and in assessing decreases in Aβ burden with therapy, we attempted to use cDNA array technology to identify patterns of altered gene expression in Aβ-overexpressing mice. This chapter documents our efforts at the analysis, interpretation, and confirmation of these efforts.

From: *Neuromethods, Vol. 37: Apoptosis Techniques and Protocols, 2nd Ed.*
Edited by: A. C. LeBlanc @ Humana Press, Inc., Totowa, NJ

2. Methods

2.1. Tissue Preparation

Three mice transgenic for the human Aβ precursor protein *(2)* and three strain-matched (C57B6/SJL) control mice, each maintained for 22 ± 2 mo (mean ± SD), were humanely killed. Mice of this age and genetic background have very robust Aβ deposition *(8)*. The mouse brains were rapidly removed, severed at the midline, and flash-frozen in liquid nitrogen.

2.2. RNA Isolation

For the analyses of gene expression in vivo, total RNA was isolated from one cortical hemisphere by the guanidinium isothiocyanate and phenol method of Chomczynski and Sacchi *(9)*. RNA was quantified by absorbance at 260 nm. The ratios of the absorbance of the RNA solutions at 260 nm vs 280 nm were 1.6–1.8.

2.3. Hybridization to Gene Array

Potential differentially expressed genes were identified by using ClonTech Atlas Mouse Expression Arrays, Version 1.2, and techniques recommended by the manufacturer (Clontech, Palo Alto, CA). Briefly, 1.3 μg of total RNA from each of the three control mouse brains, or each of the three transgenic mouse brains, was pooled, and labeled with [α-^{33}P]-dATP by using Maloney murine leukemia virus reverse transcriptase and a proprietary primer mix (Clontech). Following the labeling reaction, unincorporated radionucleotides were separated from incorporated radionucleotides by using spin-column size-separation chromatography.

Array nylon membranes were allowed to prehybridize with ExpressHyb hybridization buffer (Clontech) for 30 min at 68°C. Approximately 1,000,000 cpm of radiolabeled probe were then diluted into ExpressHyb and incubated with the nylon membranes overnight at 68°C. The membranes were washed four times with 2X saline sodium citrate (SSC), 1% SDS at 68°C for 30 min each, and then once with 0.1X SSC, 0.5% SDS and once with 2X SSC. The membranes were then exposed to phosphorimaging plates overnight. Bound radioactivity was detected and quantified by using phosphorimaging technology (FLA-2000, Fuji, Stamford, CT).

2.4. Array Analysis

Phosphorimage files were exported as .TIF files from the phosphorimaging software. These files were imported into Atlas Image 1.01, a proprietary Atlas membrane analysis software package (Clontech). The images were then overlayed with a grid. Substantial differences were noted visually between the grid overlay and the apparent placement of individual arrayed cDNAs, that is, some arrayed cDNAs were out of alignment with the overall pattern. Errors in alignment were corrected by adjusting each cell of the grid. Binding patterns between transgenic and control mice were compared by using the Atlas Image 1.01 software. Radiolabel intensities were quantified, corrected for background, and normalized for slight differences in the intensity of internal positive controls. The ratio of radioactivity bound to each cDNA between the two blots was then determined.

2.5. Reverse Transcriptase Polymerase Chain Reaction (RT-PCR)

One-microgram aliquots of RNA were subjected to reverse transcription as follows. RNA was mixed with 500 pmol of random hexamers (Roche, Indianapolis, IN) in a volume of 16 µL, incubated at 95°C for 2 min, and then placed on ice. A stock solution was then added such that the final reaction volume of 30 µL contained 200 U Superscript II (Life Technologies, Gaithersburg, MD), 500 µmol dNTPs, 40 U RNAsin, and 1X reaction buffer (Life Technologies). The solution was incubated at 20°C for 10 min, 42°C for 50 min, and the Superscript then inactivated by heating to 95°C for 2 min.

Differential gene expression was analyzed by using semiquantitative RT-PCR. In the course of this experimentation, we compared two means of detecting and quantifying PCR products, namely, incorporating radioactive dCTP during the PCR to radiolabel the PCR products or by detecting PCR products with representative members of the new generation of DNA fluorescent probes, SYBR Green (Molecular Probes, Eugene, OR). Stock PCR reaction mixtures (50 µL) were prepared on ice and contained 50 µmol dCTP, 100 µmol each of dGTP, dATP, and dTTP, 10 µCi dCTP (3000 Ci/mmol), 1.5 mmol $MgCl_2$, 1X reaction buffer (Life Technologies), 1 µmol each primer, 1 U of Taq polymerase (Life Technologies), and 1/30th of the cDNA synthesized in the RT. The stock solutions were then separated into three 14-μL aliquots that were covered with a drop of mineral oil and subjected to the indicated numbers of PCR cycles. Typical reaction conditions were 1 min at 94°C, 1 min at 55°C, and 2 min at 72°C. After amplification, the cDNAs were separated by polyacrylamide gel electrophoresis and stained with SYBR Green for 10 min as directed by the manufacturer (1/10,000 dilution in Tris-borate-EDTA buffer). The gels were then scanned by using a fluorescent scanner (FLA-2000, Fuji, Stamford, CT). After scanning, the gels were dried and subjected to phosphorimaging analysis (Fuji). In subsequent experiments, PCR products were detected and quantified by fluorescent scanning.

Sequences of the primer's used are as follows: *transcription termination factor-1* (*TTF*) sense primer, 5'GCC TCA GTG ACA GAC AGC AA 3'; *TTF* antisense primer, 5'GTT CTA ACC TCT GCA TGG CTT (155-bp product); *microtubule-associated protein-4* (*MAP-4*) sense primer, 5'GTG GGA GAA ACT GTG GAG AA; *MAP-4* antisense primer, 5'GTT GGA GAC CCT GCA GTG GG (190-bp product); *cAMP response element binding protein 1* (*CREB*) sense primer, 5'AGG AGG CCT TCC TAC AGG AA; *CREB* antisense primer, 5'ACC ATT GTT AGC CAG CTG TAT T (203-bp product); *tissue inhibitor of metalloproteinases-2* (*TIMP 2*) sense primer, 5'CTC (A/G) CT GGA CGT TGG AGG AA; *TIMP 2* antisense primer, 5'CCC ATC TGG TAC CTG TGG TT (153-bp product); *B-Raf* sense primer, 5'CAT GGT GAT GTG GCA GTG AAA; *B-Raf* antisense primer, 5'CTG GAG CCC TCA CAC CAC T (188-bp product); *numblike* sense primer, 5'CTG AGC GAT GGT TGG AGG AA; *numblike* antisense primer, 5'CAC TGG GGC AGA AAA GGG TT (149-bp product); *SLUGH* (homolog of chicken *SLUG* zinc finger protein) sense primer, 5'GGG CAC GGA CAT GAG GTA A; *SLUGH* antisense primer, 5'ACA GCA CAA GCT AAG GGC TT (189-bp product); *cyclinG* sense primer, 5'GTT TCA GAC CTG ATG AGA A; *cyclinG* antisense primer, 5'AGC ACA GAA GGC TTT GCC TT (218-bp product); *myelin proteolipid protein* (*PLP*) sense primer, 5'TGG CCA CTG GAT TGT GTT TC; *PLP* antisense primer, 5'AGG AAG AAG AAA GAG GCA GTT (164-bp product); *thymosinβ-4* (*thy B4*) sense primer, 5'CTG AGA TCG AGA AAT TCG ATA A; *thy B4* antisense primer, 5'TCA TTA CGA TTC GCC AGC TT (116-bp product); murine homolog of *disheveled* sense primer, 5'CAC AAA TGC CGT CGT CGG AA; *disheveled* antisense primer, 5'AAG TGG TGC CTC TCC ATG GG (128-bp product); *cathepsinA* sense primer, 5'TCT TCC GCC TCT TTC CGG AA; *cathepsinA* antisense primer, 5'TAG

ACC AGG GAG TTG TCG TT (161-bp product); *corticotrophin releasing hormone binding protein* (*CRHBP*) sense primer, 5'CTC TCA GAC TCC GAG TGG AA; *CRHBP* antisense primer, 5'TCC CAG CAG CTC CAC AAA GT (197-bp product); *glucose regulated protein-78* (*GRP-78*) sense primer, 5'CAC TGA TCT GCT AGA GCT GTA A; *GRP-78* antisense primer, 5'GCC ACT TGG GCT ATA GCA TT (140-bp product). All other primers were published elsewhere *(10, 11)*. Where shown, differences in gene expression were analyzed statistically by ANOVA comparison of transgenic vs wild-type mice with a post-hoc Fisher PLSD test for significance (GB-Stat, Version 6.5, Dynamic Microsystems, Silver Spring, MD).

3. Results

In these studies, we sought to distinguish quantitatively the different patterns of gene expression between a mouse model for AD and strain-matched, age-matched, wild-type mice. The gene array approach appeared to work well technically because the intensity of radiolabeling of the internal control cDNAs was well reproduced between the Aβ and control mouse samples (Fig. 1). Indeed, the vast majority of cDNAs were labeled in an essentially equivalent fashion between the two samples. To discern quantitative differences, we used Atlas 1.01 array analysis software, which identified a number of genes as differentially expressed (Table 1). These suggested differences in expression ranged as high as 17-fold.

To begin to analyze the results obtained from the cDNA arrays, we first optimized an RT-PCR assay. The advent of relatively new fluorescent DNA stains allows the fluorescent detection of PCR products at much lower levels than previous stains such as ethidium bromide. To increase sample throughput, we wished to use nonradioactive PCR quantitation as opposed to our standard radioactive PCR quantitation. The expression of several genes was examined in parallel by amplifying increasing amounts of cDNA by PCR, and then quantifying the PCR products in the same gels by either fluorescent or radioactive means. We found that the PCR product corresponding to *actin* was linear with input cDNA after 18 cycles of amplification whether quantitation was by radioactive or nonradioactive means (Fig. 2A,B). However, after 23 PCR cycles, product formation was no longer linear with input cDNA, indicating that the PCR was approaching saturation after this number of cycles and quantities of this particular cDNA. Similarly, PCR

Wild Type

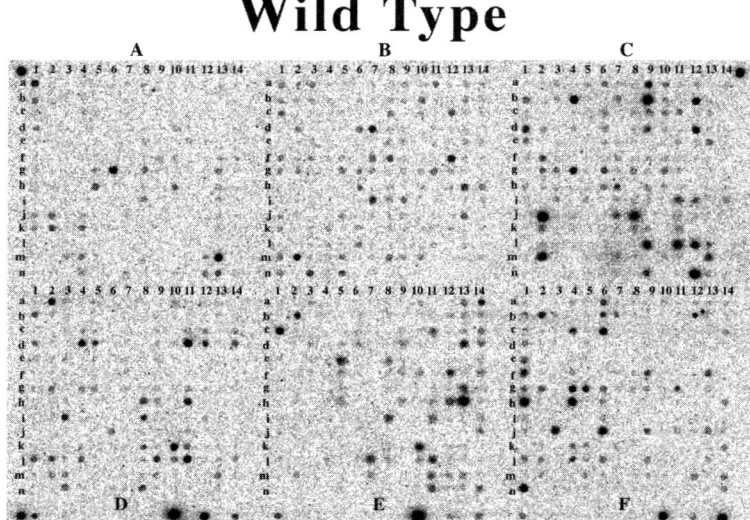

hAPP Transgenic

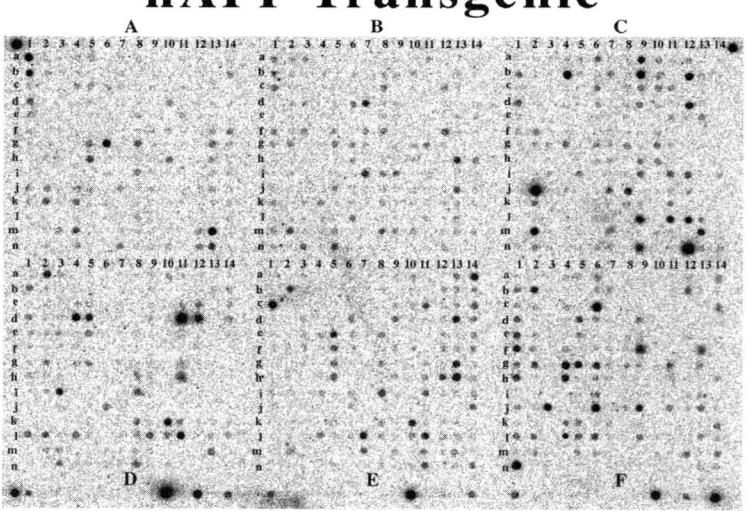

Fig. 1. Atlas mouse 1.2 cDNA arrays labeled with probes from hAPP and wild-type mouse brain RNA. Membranes spotted with approximately 1200 cDNAs were hybridized with radiolabeled cDNA probe, dried, and exposed to phosphorimaging screens. These images depict the phosphorimage results of the two cDNA arrays. Differentially expressed genes are reflected by differences in radiolabeling intensity. The bottom row of each membrane constitutes positive and negative control cDNAs, and is essentially identical between the two blots.

Table 1
Putative Differentially Expressed
Genes Identified by Atlas 1.01 Software Analysis

Grid location	Ratio APP/wild type	Gene
A01b	2.10	Myelin-oligodendrocyte glycoprotein precursor (MOG)
A14f	4.15	cAMP-responsive element-binding protein 1
B01m	0.41	Delta-like protein precursor (DLK); preadipocyte factor 1 (PREF1)
C01e	0.34	P-selectin glycoprotein ligand 1 precursor (PSGL1; SELPLG; SELP1)
C02f	0.50	Neural cadherin precursor (N-cadherin; CDH2)
C02g	0.47	Neuronal cell surface protein F3
C02n	17.27	Transcription termination factor 1 (TTF1)
C04g	0.42	Max protein. MAX PROTEIN (MYN PROTEIN)
C05i	3.51	Insulin-like growth factor binding protein 2 precursor)
C07h	0.46	B-raf proto-oncogene
C09i	2.40	Glutathione reductase
C14i	0.36	Fas l receptor; Fas antigen (Apo-1 antigen)
D03m	0.29	Angiotensin-converting enzyme (ACE) (clone ACE.5.)
D05e	5.38	Activin type I receptor
D08h	0.50	Gamma-aminobutyric acid (GABA-A) receptor, subunit
D11k	0.38	Secretogranin II precursor (SGII); chromogranin C
E12f	0.40	Mitogen-activated protein kinase p38 (MAP kinase p38)
F04c	0.41	Cathepsin B1 (CTSB)
F05d	3.42	Tissue inhibitor of metalloproteinases 2 (TIMP2)
F05l	2.81	Nonmuscle cofilin 1 (CFL1)
F06b	0.43	Alpha internexin neuronal intermediate filament protein (alpha-INX; INA)
F06d	8.38	Microtubule-associated protein 4 (MAP4; MTAP4)
F14a	4.47	Brain lipid-binding protein (BLBP)

products corresponding to *c-jun* were linear with input cDNA after 21 cycles, whether quantitation was by radioactive or nonradioactive means. However, after 26 cycles, formation of this *c-jun* PCR product again approached saturation, especially at higher amounts of added cDNA template. In summation, our experience with ethidium bromide was that PCR products were often not detected until they had reached saturation of amplification. In contrast, this newer generation of fluorescent stains with their higher sensitivity now

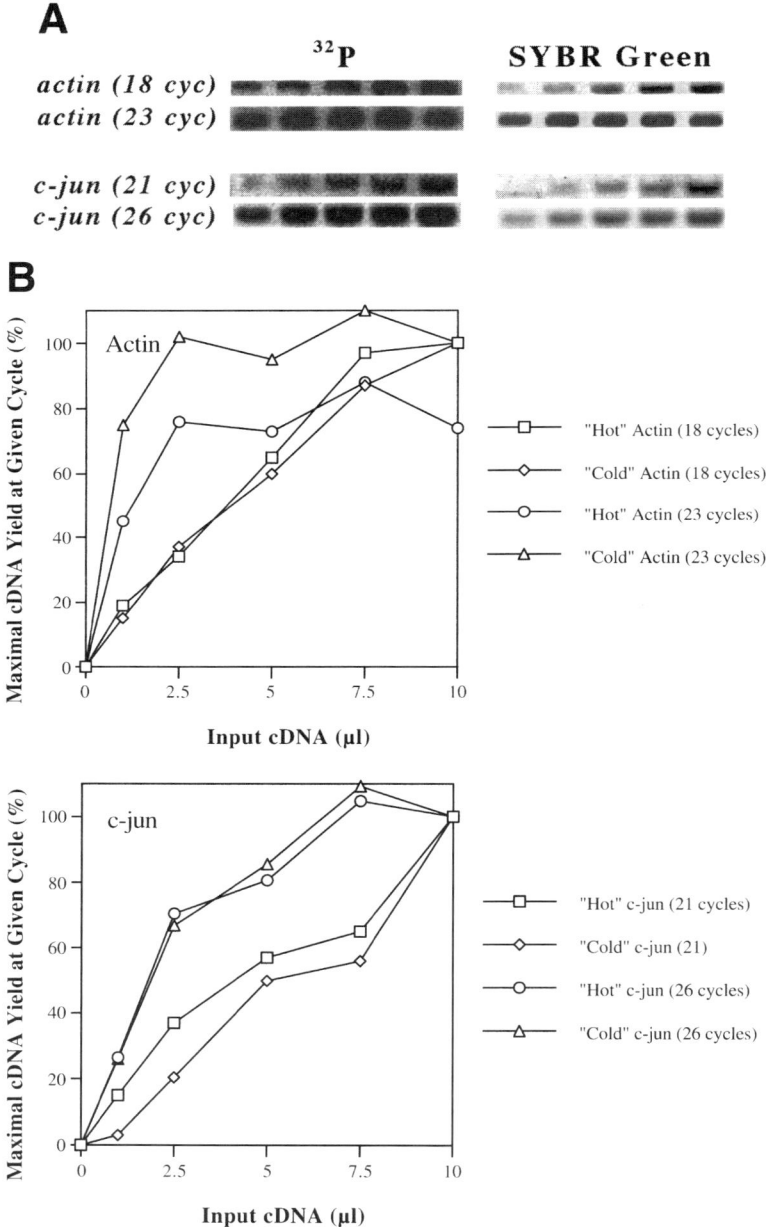

Fig. 2. Validation of nonradioactive PCR product quantitation. Increasing amounts of cDNA were amplified for either 18 and 23 cycles (*actin*) or for 21 and 26 cycles (*c-jun*). Representative gels are presented in **A**, while **B** depicts quantitation of duplicate samples.

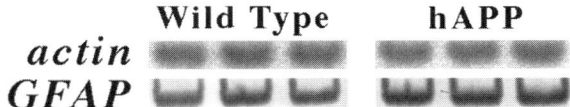

Fig. 3. Baseline patterns of gene expression in hAPP vs control mice. RNA from three hAPP transgenic mice and three age-matched, strain-matched control mice was analyzed for *actin* and *GFAP* expression by nonradioactive RT-PCR. This figure was published previously *(12)* and used here with the publisher's permission (copyright © 2000 by the Society for Neuroscience).

allows nonradioactive PCR product quantitation that is comparable to that offered by radioactive dNTP incorporation.

To validate the genetic changes suggested by the nylon membrane arrays, we began by establishing a baseline of gene expression in the mouse brain samples. We quantified the expression of *actin*, a constitutively expressed gene, as well as *GFAP*, a marker for the reactive astrocytes that are known to be present in these mice. As can be seen in Fig. 3, the expression of *actin* was unchanged with Aβ overproduction, whereas *GFAP* expression was increased approximately 80% *(12)*. We then purchased oligo pairs corresponding to a battery of selected differentially expressed mRNAs. Only the oligo pairs that produced PCR products of the appropriate size were used for quantitative studies. We then began to assess whether the genes implicated by the cDNA array were differentially expressed by performing RT-PCR studies on the pooled cDNA, that is, by comparing the mixed Aβ mouse cDNA with the mixed control mouse cDNA. This approach suggested that few, if any, of the suggested genes were actually differentially expressed (Fig. 4).

Several genes that appeared to be differentially expressed as assessed by careful visual scrutiny of the blots were not identified as such by the Atlas array analysis software (Table 2). Therefore, we decided to evaluate these genes by RT-PCR as well. Only *SLUGH* and *GRP-78* appeared to be differentially expressed as discerned by this secondary analysis (Fig. 5).

To evaluate as carefully as possible the changes in gene expression suggested by the pooled sample comparison, we performed RT-PCR on RNA from each of the individual mouse brains. Surprisingly, only one of the five RNAs suggested to be differentially

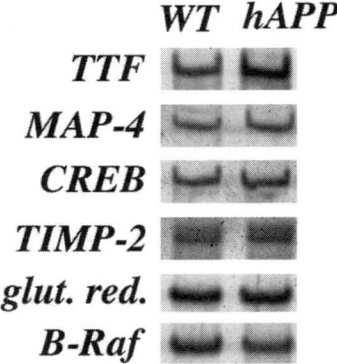

Fig. 4. Pairwise comparison of putative differentially expressed genes. Genes suggested to be differentially expressed by the Atlas 1.01 array analysis software were analyzed by nonradioactive RT-PCR. Multiple cycles of PCR were performed for each gene analysis. The number of cycles required for linear quantitation was 20–23 cycles.

Table 2
Possible Differentially Expressed
Genes Identified by Visual Inspection
of Atlas Membrane Arrays

Gene	Grid Location
numblike	A02b
slug H	A10d
cyclin G	B10b
PLP	D11d
thymosin B4	D12d
dishevelled	E06d
cathepsin A	F02l
CRHBP	E04h
GRP-78	C09d

expressed, glutathione reductase, actually appeared to be differentially expressed upon individual sample comparison. The levels of expression for the wild type and hAPP mice were 1255 ± 178 and 1945 ± 280 (mean ± SD, $n = 3$, arbitrary units), respectively, representing a 55% increase (Fig. 6).

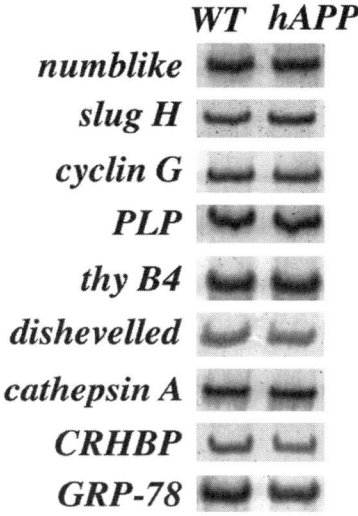

Fig. 5. Pairwise comparison of putative differentially expressed genes. Genes suggested to be differentially expressed by careful visual inspection of the Atlas Mouse 1.2 arrays phosphorimage results were analyzed by nonradioactive RT-PCR. Multiple cycles of PCR were performed for each gene analysis. The number of cycles required for linear quantitation was 20–23 cycles.

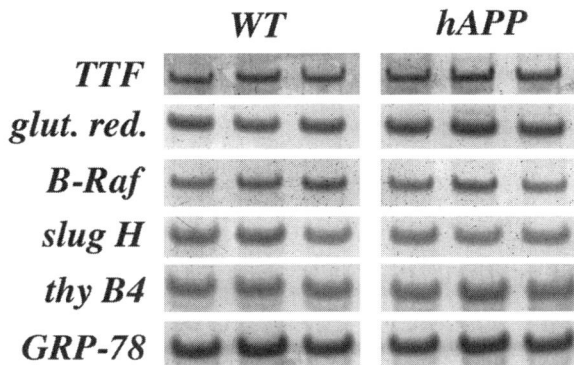

Fig. 6. Three-vs-three comparison of putative differentially expressed genes. Genes suggested to be differentially expressed from the pairwise comparisons were analyzed further by performing nonradioactive RT-PCR on each of the separate brain samples. Multiple cycles of PCR were performed for each gene analysis. The number of cycles required for linear quantitation was 20–23 cycles. ANOVA analysis of the five quantified mRNAs with a post-hoc Fishers PLSD test indicated that the increase in glutathione reductase in the hAPP mice was significant ($p < 0.05$).

4. Discussion

The primary findings described here are threefold. First, we interpret these data as suggesting that results obtained with nylon gene arrays must be confirmed by an independent methodology to discriminate between false positives and actual differences in gene expression. Second, the current generation of fluorescent DNA stains is sufficiently sensitive that nonradioactive RT-PCR produces data quantitatively similar to that of radioactive RT-PCR. Third, glutathione reductase appears increased in Aβ-overproducing mice. This result is consistent with reports of enhanced oxidative stress in these mice (13).

The availability of various gene arrays offers the suggestion that researchers can evaluate the expression of great numbers of genes simultaneously. Our results here echo those of others suggesting that gene arrays offer a strength in numbers of genes analyzed but that these data must be interpreted conservatively without a secondary confirmation (14,15).

Second, a variety of quantitative RT-PCR approaches are available currently. These differ primarily in three ways. First, "spiking" a known amount of a cRNA into an RNA mixture can provide an internal standard that mimics the mRNA of interest; comparison of the quantity of the resultant wild-type PCR product with that of the mimic can then be used to provide absolute quantitation of a particular gene (16). The strength of this approach is its absolute quantitation. The weakness is that a separate mimic must be generated for each gene of interest, which can be time-consuming if one wishes to examine a relatively large number of genes. Second, we and others have proposed the type of RT-PCR quantitation described here wherein the PCR products are separated by polyacrylamide gel electrophoresis, and the quantity of product is then quantified by either radioactive or nonradioactive means (17). The strengths of this approach is that changes in the relative expression of a given gene can be assessed in a reasonably quantitative fashion, with a minimum of cost. The primary weakness is that the problem of PCR saturation necessitates that each sample be examined at different PCR cycle numbers to ensure that one has not saturated the PCR. The third approach for quantitative RT-PCR relies on relatively new technology, wherein the quantity of double-stranded DNA present in a capillary tube PCR reaction is quantified in "real time" after each cycle by using a fluorescent double-strand-specific DNA stain and reading fluorescence through the capillary tube (reviewed in ref. *18*). The primary strength of this approach is that the linear range of PCR cycles is easily determined. The primary weakness is that nonspecific PCR product amplification is not easily distinguished from the anticipated specific PCR product. The approach can distinguish primer-dimers from product by melting the primer-dimers, but will not distinguish a specific product from a nonspecific product of grossly similar size. Hence, if a given PCR oligo pair produces a very clean PCR product, this approach will work well, but the accuracy of quantitation with this approach decreases with increasing nonspecific PCR product contamination.

The current generation of DNA stains has a greater than two orders of magnitude increased sensitivity over its predecessors, such as ethidium bromide. In our hands, we have found that PCR product generation is very frequently nonlinear by the time that bands are visible by ethidium bromide staining. This is reflected here, where a difference of five cycles made a large difference in whether the PCR products were linearly dependent on input cDNA. Five cycles of PCR would be expected to produce an approximately 10-fold increase in PCR product, given typical amplification efficiencies *(19)*.

Lastly, attributes of oxidative stress have been reported by many groups to be a hallmark of the Alzheimer's disease brain (reviewed in ref. *20*). Our finding here that glutathione reductase is increased in mice with an enhanced Aβ burden was somewhat remarkable in that we found several years ago that this same gene is also increased in human AD brains evaluated after autopsy *(11)*. The relevance of this gene induction to AD pathophysiology is unclear, but likely reflects that increased Aβ levels cause oxidative stress and the associated changes in gene expression in both humans and mice.

In conclusion, cDNA membrane arrays appear useful in that they can identify potentially differentially expressed genes. As we have found here, their primary weakness appears to be a large percentage of false positive results. Considering this weakness, we note that we have not evaluated how many real positives were missed by this technique. We also note that there are additional genes that were implicated to be differentially induced that we have not yet pursued, largely for reasons of time and priorities. Others may wish to pursue these putative differences.

References

1. Shapira, R., Austin, G. E., and Mirra, S. S. (1988) Neuritic plaque amyloid in Alzheimer's disease is highly racemized. *J. Neurochem.* **50,** 69–74.
2. Hsiao, K., Chapman, P., Nilsen, S., et al. (1996) Correlative memory deficits, Aβ elevation, and amyloid plaques in transgenic mice. *Science* **274,** 99–102.
3. Irizarry, M. C., Soriano, F., McNamara, M., et al. (1997) Abeta deposition is associated with neuropil changes, but not with overt neuronal loss in the human amyloid precursor protein V717F (PDAPP) transgenic mouse. *J. Neurosci.* **17,** 7053–7059.
4. Frautschy, S. A., Yang, F., Irrizarry, M., et al. (1998) Microglial response to amyloid plaques in APPsw transgenic mice. *Am. J. Pathol.* **152,** 307–317.
5. Schenk, D., Barbour, R., Dunn, W., et al. (1999) Immunization with amyloid-beta attenuates Alzheimer-disease-like pathology in the PDAPP mouse. *Nature* **400,** 173–177.
6. Morgan, D., Diamond, D. M., Gottschall, P. E., et al. (2000) A beta peptide vaccination prevents memory loss in an animal model of Alzheimer's disease. *Nature* **408**(6815), 982–985.
7. Janus, C., Pearson, J., McLaurin, J., et al. (2000) A beta peptide immunization reduces behavioural impairment and plaques in a model of Alzheimer's disease. *Nature* **408**(6815), 979–982.
8. Kawarabayashi, T., Younkin, L., Saido, T., Shoji, M., Ashe, K., and Younkin, S. (2001) Age-dependent changes in brain, CSF, and plasma amyloid (beta) protein in the Tg2576 transgenic mouse model of Alzheimer's disease. *J. Neurosci.* **21**(2), 372–381.
9. Chomczynski, P. and Sacchi, N. (1987) Single-step method of RNA isolation by acid guanidinium thiocynate-phenol-chloroform extraction. *Anal. Biochem.* **162,** 156–159.
10. Estus, S., Tucker, H. M., van Rooyen, C., et al. (1997) Aggregated amyloid-β protein induces cortical neuronal apoptosis and concomitant "apoptotic" pattern of gene induction. *J. Neurosci.* **17,** 7736–7745.
11. Aksenov, M. Y., Tucker, H. M., Nair, P., et al. (1998) The expression of key oxidative stress-handling genes in different brain regions in Alzheimer's disease. *J. Mol. Neurosci.* **11,** 151–164.
12. Tucker, H. M., Kihiko, M. E., Caldwell, J. N., et al. (2000) Amyloid-beta activates and is degraded by the plasmin system. *J. Neurosci.* **20,** 3937–3946.
13. Smith, M. A., Hirai, K., Hsiao, K., et al. (1998) Amyloid-beta deposition in Alzheimer transgenic mice is associated with oxidative stress. *J. Neurochem.* **70,** 2212–2215.
14. Claverie, J. M. (1999) Computational methods for the identification of differential and coordinated gene expression. *Hum. Mol. Genet.* **8**(10), 1821–1832.
15. Miller, R. A., Galecki, A., and Shmookler-Reis, R. J. (2001) Interpretation, design, and analysis of gene array expression experiments. *J. Gerontol. A. Biol. Sci. Med. Sci.* **56**(2), B52–B57.
16. Karge, W. H. III, Schaefer, E. J., and Ordovas, J. M. (1998) Quantification of mRNA by polymerase chain reaction (PCR) using an internal standard and a nonradioactive detection method. *Meth. Mol. Biol.* **110,** 43–61.

17. Estus, S. (1997) Optimization and validation of RT-PCR as a tool to analyze apoptotic gene expression, in *Neuromethods 29: Apoptosis Techniques and Protocols* (Poirier, J., ed.), Humana Press, Totowa, NJ, pp. 67–84.
18. Jung, R., Soondrum, K., and Neumaier, M. (2000) Quantitative PCR. *Clin. Chem. Lab. Med.* **38**(9), 833–836.
19. Golde, T. E., Estus, S., Usiak, M., Younkin, L. H., and Younkin, S.G. (1990) Expression of beta amyloid protein precursor mRNAs: recognition of a novel alternatively spliced form and quantitation in Alzheimer's disease using PCR. *Neuron* **4**(2), 253–267.
20. Smith, M. A., Nunomura, A., Zhu, X., Takeda, A., and Perry, G. (2000) Metabolic, metallic, and mitotic sources of oxidative stress in Alzheimer disease. *Antioxid. Redox. Signal* **2**(3), 413–420.

Glyceraldehyde-3-phosphate Dehydrogenase as a Target for Antiapoptotic Drugs

Mark D. Berry and Paula C. Ashe

1. Introduction

Glyceraldehyde-3-phosphate dehydrogenase (GAPDH; EC 1.2. 1.12) has classically been regarded as an ubiquitous enzyme of little importance beyond its role in glycolysis. Indeed, the most frequent reference to GAPDH in recent scientific literature is as the "house-keeping" gene used to standardize Northern blots. A careful examination of the literature of the last 25 years, however, reveals a number of novel actions of GAPDH beyond a role in glycolysis. A review of these novel functions of GAPDH is beyond the scope of this chapter, and the reader is referred to a number of recent thorough reviews by Michael Sirover for further details *(1,2)*. Of the novel functions identified for GAPDH, one of the most intriguing is as an integral part of one or more apoptotic cascades. Here, we will briefly outline the evidence for GAPDH not only playing a role in the initiation of apoptosis, but also being a target of a number of known antiapoptotic compounds. A more in-depth discussion of the role of GAPDH in apoptosis is provided in our recent review article *(3)*.

Oxidative stress is one of the best documented inducers of apoptosis *(4)*. In isolated rabbit aorta, an increase in GAPDH mRNA has been reported in response to oxidative stress *(5)*. Similarly, hypoxia/ischemia is a popular model for the study of apoptosis and may in part be a result of increased oxidative stress. Although the levels

From: *Neuromethods, Vol. 37: Apoptosis Techniques and Protocols, 2nd Ed.*
Edited by: A. C. LeBlanc @ Humana Press, Inc., Totowa, NJ

of apoptosis were not studied, increases in GAPDH mRNA have been observed in response to hypoxia/ischaemia both in vivo (6) and in vitro (7). Whether these changes in GAPDH mRNA result in a similar increase in GAPDH protein is unknown. While Ito et al. (5) failed to detect any change in glycolytic activity, this does not preclude an increase in GAPDH protein. A number of the novel functions of GAPDH have been shown to be independent of glycolytic activity (2), and multiple GAPDH isoforms (8–10) and mRNA species (11) have been sporadically reported. This aspect of GAPDH regulation has, however, been poorly studied.

Hindlimb unweighting has been used as a model system of prolonged muscular inactivity. Under such circumstances skeletal muscle atrophy occurs, and this has been shown to be, at least in part, due to apoptosis (12). Following hindlimb unweighting an increase in GAPDH mRNA is seen in the unweighted skeletal muscle. Interestingly, hibernating animals, which during hibernation are in a prolonged state of muscular inactivity, do not show the degree of skeletal muscle atrophy that one would expect following a similar degree of muscle disuse in a euthermic animal. Whether this represents the activation of an endogenous antiapoptotic programme is unknown. Consistent with such an effect, however, hibernating animals show a marked resistance to both cerebral ischemia (13) and traumatic brain injury (14) in comparison to their euthermic counterparts, an effect that is not a function of alterations in glutamate release (14) as had previously been hypothesized. Interestingly, it has been shown that in at least one hibernating species, there is a decrease in both GAPDH mRNA and protein in skeletal muscle during hibernation (9). While this may simply represent a downregulation of glycolytic activity, it is interesting to note that this loss of GAPDH protein was not observed throughout the animal. Decreases were observed in tissues that could reasonably be expected to be susceptible to undergoing apoptosis in a hibernating animal, such as skeletal muscle and the heart. In contrast, no change was observed in tissues such as the liver and lungs (15).

In summary, although circumstantial, there is evidence that in cases where an increase in apoptosis occurs, this may be associated with an increase in GAPDH. In contrast, in conditions where an endogenous antiapoptotic programme may be required, GAPDH levels may decrease. The above results are summarized in Table 1.

During recent years direct evidence has shown GAPDH to be involved in the initiation of apoptosis. Initially, GAPDH mRNA

Table 1
Changes in GAPDH Protein
and mRNA Associated with Changes in Apoptosis

Stressor	Apoptotic response	GAPDH response
Oxidative stress	Increased apoptosis	Increased mRNA
Hypoxia/ischaemia	Increased apoptosis	Increased mRNA
Hindlimb unweighting	Increased apoptosis	Increased mRNA
Hibernation	Decreased apoptosis	Decreased mRNA and protein

See text for references.

Table 2
Relationship of GAPDH
to Neurodegenerative Disease-Related Proteins

Protein	Disease	Apoptosis?	GAPDH involvement
β-Amyloid precursor	Alzheimer's	Yes	Binds
Huntingtin	Huntington's	Yes	Binds expanded CAG repeat
Numerous (e.g., ataxin 1)	Spinocerebellar ataxias	Unknown	Binds expanded CAG repeat
—	Parkinson's	Yes	Increased nuclear location

For references see text.

and protein was shown to be increased following initiation of apoptosis in cerebellar granule cell cultures by lowering the extracellular potassium concentration *(16)*. These increases in GAPDH protein were associated with the particulate fraction. GAPDH antisense oligonucleotide prevents changes in GAPDH and induction of apoptosis *(16)*. In subsequent years, similar changes in GAPDH were reported in response to aging-induced apoptosis *(17)* and ara-C–induced apoptosis *(18,19)*. Further, GAPDH changes during the initiation of apoptosis have also been observed in cerebro-cortical neurons *(20)* and in non-neuronal cells such as macrophages *(21)* and thymocytes *(22)*. The increases in particulate GAPDH are now known to be specifically associated with an increase in nuclear GAPDH *(10,23,24)*, and are independent of glycolytic activity *(25)*.

The possible relevance of GAPDH changes to human neurodegenerative disorders has recently begun to be appreciated. As summarized in Table 2, GAPDH has been shown to interact with

Table 3
Antiapoptotic Compounds
and GAPDH-Binding Affinity

Compound	Antiapoptotic?	GAPDH-binding affinity
R-deprenyl	Yes	Low nM
CGP 3466	Yes	Low nM
17β-estradiol	Yes	Low nM
R-2HMP	Yes	Unknown

For references see text.

a number of proteins implicated in human neurodegenerative disorders. As long ago as 1993, GAPDH was shown to interact with the β-amyloid precursor protein *(26)*. Further, antibodies raised against Alzheimer's disease β-amyloid plaques have been shown to identify GAPDH *(16)*. Better documented is the binding of GAPDH to proteins containing polyglutamine expansions *(27–29)*. A number of these proteins have now been identified as the cause of several degenerative disorders *(30)*. Intriguingly, the toxicity of these mutated proteins appears to be associated with the appearance of the polyglutamine region in the nucleus. This is the region to which GAPDH binds, and as described above, the nuclear appearance of GAPDH appears to initiate apoptosis. Recently, an increase in nuclear GAPDH has been reported in postmortem Parkinson's disease substantia nigra *(31)*. Further, this increase is associated with the melanized neurons, which are particularly susceptible to degeneration in Parkinson's disease. Apoptosis has been implicated in all of these disorders *(31–35)*.

The mechanism by which GAPDH appearance in the nucleus initiates apoptosis is unknown, although this does appear to be independent of glycolytic activity *(25)*. The nonglycolytic functions of GAPDH that may be involved in initiating apoptosis have been described previously *(3)* and will not be further discussed here.

Four known antiapoptotic compounds have either been shown or implied to bind to GAPDH. These data are summarized in Table 3, and will be discussed in more detail in the following sections. In each case, it is striking to note that antiapoptotic efficacy appears to correlate with affinity for GAPDH binding.

2. Methods

2.1. Antisense Technology

We have adopted a method similar to that described by Ishitani et al. *(17,20)* in order to show directly the requirement of new GAPDH synthesis for apoptosis to occur. In these experiments we have utilized our previously described *(36,37)* cultured cerebellar granule cell protocol, although the method has also been applied to a number of other cell types *(20,22)*. Antisense and sense sequences were based on those published by Ishitani et al. *(17,20)* and designed to be complementary to the sequence surrounding the ATG start codon. Oligonucleotides designed to interact with other coding sequences of GAPDH mRNA have also been shown to be active *(17,20,22)*. The sequences we utilized were 5'-G*ACCTTCACCA TCTTGTCT*A-3' for the antisense and the inverse sequence for the sense (with phosphorothioate linkages in the equivalent positions). Asterisks indicate the location of phosphorothioate linkages. The use of a sense sequence as an internal control for non-specific effects of the antisense sequence has been debated *(38)*. A more appropriate control may be a scrambled version of the antisense sequence, which would have the same nucleotide composition as the antisense sequence but in a random (non-sense) sequence. Use of both a sense and a non-sense control are not uncommon. Further, the position and number of phosphorothioate linkages is an important consideration. We have found that sequences containing more phosphorothioate linkages than above are toxic (Zhang, Berry, and Paterson, unpublished observations). It has been reported that a single phosphorothioate linkage confers as much protection from nuclease activity as do multiple linkages.

Oligonucleotide sequences are obtained as a lyophilized powder. This powder is resuspended in distilled water to a stock concentration of 1 mM. These stock solutions are stored frozen at −20°C. Before use, the stock solution is thawed and diluted as required with distilled water. Twenty-four hours following plating of cerebellar granule cells, antisense or sense oligonucleotides are added 90 min before feeding the cells. Addition of oligonucleotides in 2- to 5-μL aliquots at active concentrations is sufficient. No active transfection protocol is required, making this a highly attractive method. Previously, it has been reported that primary cell cultures in general uptake oligonucleotides very well *(38)*.

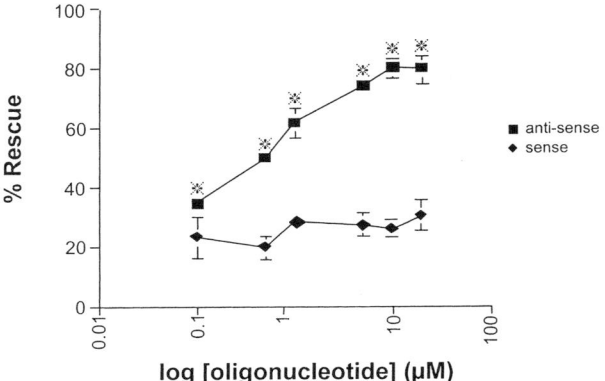

Fig. 1. Dose–response curve for antisense oligonucleotides to GAPDH. Oligonucleotides were incubated with cultured cerebellar granule cells at the doses indicated for 90 min before addition of either neuron culture medium and induction of apoptosis with cytosine arabinoside (300 μ*M*), or addition of low-K⁺ neuron culture media. Cells were maintained in culture for a further 24 h and were then fixed, and stained with *bis*-benzimide. Normal and antiapoptotic cells were counted under a fluorescent microscope.

We have studied the effects of such treatments on apoptosis induced in one of two ways, either by feeding the cells with a low-K⁺ medium, or immediately following feeding with regular medium, addition of cytosine-β-D-arabinofuranoside (Ara-C) at a concentration of 300 μM. Using this protocol, we have demonstrated an almost complete protection against apoptosis initiated by ara-C following antisense treatment *(39)*. No such effect was seen in low-K⁺-induced apoptosis, while in both cases the sense sequence was without effect *(39)*. This effect of antisense pretreatment is dose-dependent in nature (Fig. 1), showing a maximum effect at a concentration of 10 μM. This is in good agreement with results reported elsewhere *(17)*. Experiments such as these provide excellent evidence of a causative role of new GAPDH synthesis in apoptosis. While GAPDH antisense has been reported to prevent GAPDH translocation to the nucleus (*see*, e.g., ref. *23*), what has actually been demonstrated is a prevention of the increase of GAPDH protein in the nucleus. While the exact mechanism of action of antisense treatments is a matter of some debate, there is no known proposal whereby an antisense sequence would pre-

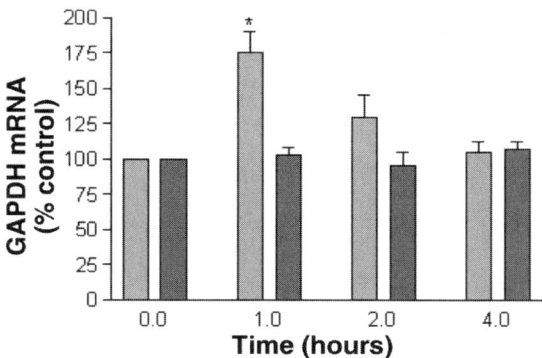

Fig. 2. Time-course of GAPDH mRNA changes following Ara-C-induced (*light tint*) and low-K$^+$-induced (*dark tint*) apoptosis. At various time points following the induction of apoptosis, cells were fixed and collected. Standard Northern blot techniques were employed using a commercially available GAPDH probe (Clontech, Palo Alto, CA). Northern blots were quantified using ImageQuant software.

vent movement of an existing protein from one cellular compartment to another (*see* ref. *40*). What would appear to be the most feasible explanation is that there is a prevention of synthesis of nuclear GAPDH. This nuclear GAPDH may in fact be a specific GAPDH isozyme (*10*). In agreement with this, we have shown a transient increase in GAPDH mRNA 1 h following Ara-C treatment, which is not seen following low-K$^+$-induced apoptosis (Fig. 2).

2.2. Green Fluorescent Protein-Tagged GAPDH

While antisense knockdown studies argue convincingly for a new GAPDH protein synthesis being required for certain forms of apoptosis to occur, there are also convincing studies that at least some degree of translocation occurs. Shashidharan and co-workers (*24*) have reported on the construction of a GFP-tagged GAPDH. Use of such a system offers considerable advantage over those previously used in that, in combination with confocal microscopy, it allows a direct determination of whether GAPDH protein is indeed translocating from one cellular compartment to another, or whether the increases observed in cellular compartments represent merely an increase in new GAPDH synthesis that is targeted to a specific compartment. Using this system, it has been clearly demonstrated that a small proportion of cells exhibits translocation of the GFP tagged GAPDH following induction of apoptosis. The number of cells exhibiting such an effect is somewhat limited, however, with a maximum of 10% of cells showing GAPDH translocation *(24)*. While the number of apoptotic cells was not reported, it seems unlikely that there would only be a 10% level of apoptosis following treatment with H_2O_2. The discrepancy between the number of cells with GAPDH translocation and the number of apoptotic cells requires clarification. Of possible relevance here is the recent report that while N-terminal GFP-tagged GAPDH shows inclusion formation, C-terminal tagged does not *(41)*. The GFP-GAPDH construct used by Shashidharan et al. was C-terminal linked. This clearly demonstrates the care required in designing GFP-tagged proteins. In the design of such studies, consideration should be given to the presence and location of subcellular organelle targeting sequences within the protein of interest. In the absence of such sequences, as may be the case with GAPDH (*see* refs. *1* and *2*), the regions of the protein required for binding to a chaperone should be considered.

2.3. Identification of GAPDH as a Binding Site

The demonstration that GAPDH antisense pretreatment selectively prevents at least some forms of apoptosis, raised the possibility that GAPDH may be a potential target for antiapoptotic drugs. The interaction between GAPDH and antiapoptotic drugs has been tested by two main methods: affinity binding *(42)* and size-exclusion chromatography *(43)*. CGP3466 and desmethyldeprenyl were immobilized on Toyopearl AF amino 650 M resin as described by Zimmerman et al. *(44)*. Extracts from rat hippocampi were incubated with the drug-immobilized resin to determine which cellular proteins interact with these known antiapoptotic compounds *(42)*. The washed resin containing the bound proteins was resuspended in SDS-PAGE sample buffer and electrophoresed on a 12.5% gel. Proteins were stained with a nonfixing silver stain and identified by nanoelectrospray mass spectrometry. GAPDH was identified as one of the proteins affinity-precipitated by the immobilized CGP3466. This interaction was determined to be specific by the lack of binding to both underivatized Toyopearl resin and to acetylated Toyopearl resin. Further proof of the specificity of binding between CGP3466 and GAPDH came from the demonstration that purified rabbit muscle GAPDH also bound to the immobilized CGP3466 but not to underivatized or acetylated Toyopearl resin. The binding of purified GAPDH to the resin also dem-onstrated that the association between CGP3466 and GAPDH was direct and not mediated by other protein interactions. Protein bind-ing to immobilized desmethyldeprenyl was not described except to note that the binding to this compound was not as selective as that to CGP3466. To demonstrate further the interaction between CGP3466 and GAPDH, photoaffinity labeling of CGP3466 was performed to assess interactions in solution between the drug and cellular proteins. GAPDH was demonstrated to be the only pro-tein that was consistently present in both the affinity precipita-tion and affinity labeling experiments thus supporting an important role for GAPDH in the mechanism of antiapoptotic compounds.

Photoaffinity labeling of CGP3466 has also been performed by Carlile et al. *(43)*, who demonstrated specific binding to GAPDH in extracts isolated from partially neuronally differentiated PC12 cells. Examination of the tetrameric structure of GAPDH, the favored form in solution, led to the conclusion that CGP3466 may bind in the central channel of the GAPDH tetramer. Incubation of cells with a polyclonal sheep antibody to GAPDH that blocks entry to the channel demonstrated a reduction in the intensity of fluorescently labeled CGP3466 as compared to that seen in cells not preincubated with the antibody. A monoclonal antibody to GAPDH also did not affect the intensity of CGP3466 fluorescence. To investigate further the interaction between CGP3466 and GAPDH, Carlile et al. investigated the oligomeric structure of GAPDH in the presence of CGP3466 and desmethyldeprenyl by size-exclu-sion chromatography. The results indicated that both CGP3466 and desmethyldeprenyl resulted in a conversion of the tetrameric form of GAPDH to a dimeric form. In addition, CGP3466 and des-methyldeprenyl also appeared to displace RNA from tetrameric GAPDH and reduce the GAPDH to its dimeric constituents. This is of particular interest since it has previously been reported that free-radical species, which are well-documented inducers of apop-tosis, displace tetrameric GAPDH from its cytoplasmic RNA bind-ing sites *(45,46)*. These results suggest that GAPDH is indeed a major target of antiapoptotic compounds and that stabilization of GAPDH in a dimeric form may prevent its translocation to the nucleus during apoptosis. This explanation, however, does not account for the increase in GAPDH synthesis reported to occur during apoptosis nor for the ability of antisense oligonucleotides to prevent apoptosis. The possibility exists that translocation of

GAPDH to the nucleus is necessary to initiate the production of new GAPDH. In making this conclusion, however, one must recognize that the amount of GAPDH required in the nucleus would be exceptionally small. This is evidenced by the demonstration that the nuclear appearance of the GAPDH protein occurs following the induction of mRNA which is reported to be maximal one hour following the induction of apoptosis (Fig. 2; 47).

There is therefore very good evidence that at least some antiapoptotic compounds can bind to GAPDH. Further, it has been reported that the known antiapoptotic compound 17β-estradiol *(48,49)* also binds with high affinity to GAPDH *(50)*. Intriguingly, the affinity for the interaction with GAPDH was in the low-nanomolar region, an affinity strikingly similar to that reported for the antiapoptotic activity of 17β-estradiol. In addition to 17β-estradiol, R-2HMP and its related derivatives have been shown to be antiapoptotic due to an action at the same site of action as R-deprenyl *(36)* and CGP 3466 *(39)*. We are currently applying affinity chromatography techniques to investigate directly the nature of the interaction of known antiapoptotic compounds with GAPDH. The recent advent of BIAcore technology will further assist in the characterization of GAPDH, and other binding sites, as targets for antiapoptotic compounds.

References

1. Sirover, M. A. (1997) Role of glycolytic protein, glyceraldehyde-3-phosphate dehydrogenase, in normal cell function and in cell pathology. *J. Cell. Biochem.* **66**, 133–140.
2. Sirover, M. A. (1999) New insights into an old protein: the functional diversity of mammalian glyceraldehyde-3-phosphate dehydrogenase. *Biochim. Biophys. Acta* **1432**, 159–184.
3. Berry, M. D. and Boulton, A. A. (2000) Glyceraldehyde-3-phosphate dehydrogenase and apoptosis. *J. Neurosci. Res.* **60**, 150–154.
4. Tong, L., Toliver-Kinsky, T., Taglialatela, G., Werrbach-Perez, K., Wood, T., and Perez-Polo, J. R. (1998) Signal transduction in neuronal death. *J. Neurochem.* **71**, 447–459.
5. Ito, Y., Pagano, P. J., Tornheim, K., Brecher, P., and Cohen, R. A. (1996) Oxidative stress increases glyceraldehyde-3-phosphate dehydrogenase mRNA levels in isolated rabbit aorta. *Am. J. Physiol.* **270**, H81–H87.
6. Feldhaus, L. M. and Liedtke, A. J. (1998) mRNA expression of glycolytic enzymes and glucose transporter proteins in ischemic myocardium with and without reperfusion. *J. Mol. Cell Cardiol.* **30**, 2475–2485.
7. Graven, K. K. and Farber, H. W. (1998) Endothelial cell hypoxic stress proteins. *J. Lab. Clin. Med.* **132**, 456–463.

8. Edwards, Y. H., Clark, P., and Harros, H. (1976) Isozymes of glyceralde-hyde-3-phosphate dehydrogenase in man and other mammals. *Ann. Hum. Genet.* **40,** 67–77.

9. Soukri, A., Valverde, F., Hafid, N., Elkebbaj, M. S., and Serrano, A. (1995) Characterization of muscle glyceraldehyde-3-phosphate dehydrogenase isoforms from euthermic and induced hibernating *Jaculus orientalis. Biochim. Biophys. Acta* **1243,** 161–168.

10. Saunders, P. A. Chen, R.-W., and Chuang, D.-M. (1999) Nuclear transloca-tion of glyceraldehyde-3-phosphate dehydrogenase isoforms during neu-ronal apoptosis. *J. Neurochem.* **72,** 925–932.

11. Piechaczyk, M., Blancard, J. M., Riad-El Sabouty, S., Dani, C., Marty, L., and Jeanteur, P. (1984) Unusual abundance of vertebrate 3-phosphate dehy-drogenase pseudogenes. *Nature* **312,** 469–471.

12. Allen, D. L., Linderman, J. K., Roy, R. R., et al. (1997) Apoptosis: a mecha-nism contributing to remodelling of skeletal muscle in reponse to hindlimb unweighting. *Am. J. Physiol.* **273,** C579–C587.

13. Frerichs, K. U. and Hallenbeck, J. M. (1998) Hibernation in ground squir-rels induces state and species-specific tolerance to hypoxia and aglyce-mia: an in vitro study in hippocampal slices. *J. Cereb. Blood Flow Metab.* **18,** 168–175.

14. Drew, K. L., Zhou, F., Hu, Y., and Braddock, J. F. (2000) Microbial origin of glutamate in microdialysis studies of hibernating ground squirrels. *Soc. Neurosci. Abstr.* **26,** 188.2.

15. Soukri, A., Valverde, F., Hafid, N., Elkebbaj, M. S., and Serrano, A. (1996) Occurrence of a differential expression of the glyceraldehyde-3-phosphate dehydrogenase gene in muscle and liver from euthermic and induced hiber-nating jerboa. *Gene* **181,** 139–145.

16. Sunaga, K., Takahashi, H., Chuang, D.-M., and Ishitani, R. (1995) Glyceral-dehyde-3-phosphate dehydrogenase is over-expressed during apoptotic death of neuronal cultures and is recognised by a monoclonal antibody against amyloid plaques from Alzheimer's brain. *Neurosci. Lett.* **200,** 133–136.

17. Ishitani, R., Sunaga, K., Hirano, A., Saunders, P., Katsube, N., and Chuang, D.-M. (1996) Evidence that glyceraldehyde-3-phosphate dehydrogenase is involved in age-induced apoptosis in mature cerebellar neurons in culture. *J. Neurochem.* **66,** 928–935.

18. Ishitani, R. and Chuang, D.-M. (1996) Glyceraldehyde-3-phosphate dehy-drogenase antisense oligodeoxynucleotides protect against cytosine arabi-noside-induced apoptosis in cultured cerebellar neurons. *Proc. Natl. Acad. Sci. USA* **93,** 9937–9941.

19. Zhang, D., Berry, M. D., Paterson, I. A., and Boulton, A. A. (1999) Two possible sites of action of R-2HMP: changes of mitochondrial membrane potential and sub-cellular GAPDH protein. *Soc. Neurosci. Abstr.* **25,** 606.18.

20. Ishitani, R., Kimura, M., Sunaga, K., Katsube, N., Tanaka, M., and Chuang, D.-M. (1996) An antisense oligodeoxynucleotide to glyceraldehyde-3-phos-phate dehydrogenase blocks age-induced apoptosis of mature cerebrocor-tical neurons in culture. *J. Pharmacol. Exp. Ther.* **278,** 447–454.

21. Messmer, U. K. and Brune, B. (1996) Modification of macrophage glyceral-dehyde-3-phosphate dehydrogenase in response to nitric oxide. *Eur. J. Phar-macol.* **302,** 171–182.

22. Sawa, A., Kahn, A. A., Hester, L. D., and Snyder, S. H. (1997) Glyceralde-hyde-3-phosphate dehydrogenase: nuclear translocation participates in neu-ronal and nonneuronal cell death. *Proc. Natl. Acad. Sci. USA* **94,** 11,669–11,674.

23. Ishitani, R., Tanaka, M., Sunaga, K., Katsube, N., and Chuang, D.-M. (1998) Nuclear localization of overexpressed glyceraldehyde-3-phosphate dehy-drogenase in cultured cerebellar neurons undergoing apoptosis. *Mol. Pharmacol.* **53,** 701–707.

24. Shashidharan, P., Chalmers-Redman, R. M. E., Carlile, G. W., et al. (1999) Nuclear translocation of GAPDH-GFP fusion protein during apoptosis. *Neuro-Report* **10,** 1149–1153.

25. Saunders, P. A., Chalecka-Franaszek, E., and Chuang, D.-M. (1997) Sub-cellular distribution of glyceraldehyde-3-phosphate dehydrogenase in cere-bellar granule cells undergoing cytosine arabinoside-induced apoptosis. *J. Neurochem.* **69,** 1820–1828.

26. Schulze, H., Schuler, A., Stuber, D., Dobeli, H., Langen, H., and Huber, G. (1993) Rat brain glyceraldehyde-3-phosphate dehydrogenase interacts with the recombinant cytoplasmic domain of Alzheimer's β-amyloid precursor protein. *J. Neurochem.* **60,** 1915–1922.

27. Burke, J. R., Enghild, J. J., Martin, M. E., et al. (1996) Huntingtin and DRLPA proteins selectively interact with the enzyme GAPDH. *Nat. Med.* **2,** 347–350.

28. Gentile, V., Sepe, C., Calvani, M., et al. (1998) Tissue transglutaminase-catalyzed formation of high–molecular-weight aggregates *in vitro* is favored with long polyglutamine domains: a possible mechanism contributing to CAG-triplet diseases. *Arch. Biochem. Biophys.* **352,** 314–321.

29. Koshy, B., Matilla, T., Burright, E. N., et al. (1996) Spinocerebellar ataxia type-1 and spinobulbar muscular atrophy gene products interact with glyc-eraldehyde-3-phosphate dehydrogenase. *Hum. Mol. Genet.* **5,** 1311–1318.

30. Koshy, B. and Zoghbi, H. Y. (1997) The CAG/polyglutamine tract diseases: gene products and molecular pathogenesis. *Brain Pathol.* **7,** 927–942.

31. Tatton, N. A. (2000) Increased caspase 3 and bax immunoreactivity accom-pany nuclear GAPDH translocation and neuronal apoptosis in Parkinson's disease. *Exp. Neurol.* **166,** 29–43.

32. Portera-Cailliau, C., Hedreen, J. C., Price, D. L., and Koliatsos, V. E. (1995) Evidence for apoptotic cell death in Huntington disease and excitotoxic animal models. *J. Neurosci.* **15,** 3775–3787.

33. Thomas, L. B., Gates, D. J., Richfield, E. K., O'Brien, T. F., Schweitzer, J. B., and Steindler, D. A. (1995) DNA end labelling (TUNEL) in Huntington's disease and other neuropathological conditions. *Exp. Neurol.* **133,** 265–272.

34. Dragunow, M., Faull, R. L. M., Lawlor, P., et al. (1995) *In situ* evidence for DNA fragmentation in Huntington's disease striatum and Alzheimer's dis-ease temporal lobes. *NeuroReport* **6,** 1053–1057.

35. Lassmann, H., Bancher, C., Breitschopf, J., et al. (1995) Cell death in Alz-heimer's disease evaluated by DNA fragmentation in situ. *Acta Neuropathol.* **89,** 35–41.

36. Paterson, I. A., Zhang, D., Warrington, R. C., and Boulton, A. A. (1998) *R*-Deprenyl and *R*-2-heptyl-*N*-methylpropargylamine prevent apoptosis in cerebellar granule neurons induced by cytosine arabinoside but not low extra-cellular potassium. *J. Neurochem.* **70,** 515–523.

37. Berry, M. D. (1999) N^8-acetyl spermidine protects rat cerebellar granule cells from low K$^+$-induced apoptosis. *J. Neurosci. Res.* **55,** 341–351.

38. Leslie, R. and James, D. (1997) Antisense gene knockdown in the nervous system revisited: optimism for the future. *Trends Neurol. Sci.* **20,** 321–322.
39. Berry, M. D. (1999) R-2HMP, an orally active agent combining independent anti-apoptotic and MAO-B inhibitory activities. *CNS Drug Rev.* **5,** 105–124.
40. Oberbauer, R. (1997) Not nonsense but antisense—applications of antisense oligonucleotides in different fields of medicine. *Wien. Klin. Wochen.* **109,** 40–46.
41. Ishitani, R., Tatton, N. A., Tsuchiya, K., Yamada, M., and Takahashi, H. (2000) Nuclear accumulation of glyceraldehyde-3-phosphate dehydrogenase (GAPDH) in Machado-Joseph disease. *Soc. Neurosci. Abstr.* **26,** 225.13.
42. Kragten, E., Lalande, I., Zimmermann, K., et al. (1998) Glyceraldehyde-3-phosphate dehydrogenase, the putative target of the antiapoptotic compounds CGP 3466 and R-(–)-deprenyl. *J. Biol. Chem.* **273,** 5821–5828.
43. Carlile, G. W., Chalmers-Redman, R. M. E., Tatton, N. A., Pong, A., Borden, K. E., and Tatton, W. G. (2000) Reduced apoptosis after nerve growth factor and serum withdrawl: conversion of tetrameric glyceraldehyde-3-phosphate dehydrogenase to a dimer. *Mol. Pharmacol.* **57,** 2–12.
44. Zimmermann, K., Roggo, S., Kragten, E., Fürst, P., and Waldmeier, P. (1998) Synthesis of tools for target identification of the anti-apoptotic compound CGP 3466; part I. *Bioorg. Med. Chem. Lett.* **8,** 1195–1200.
45. Marin, P., Maus, M., Bockaert, J., Glowinski, J., and Premont, J. (1995) Oxygen free radicals enhance the nitric oxide-induced covalent NAD(+)-linkage to neuronal glyceraldehyde-3-phosphate dehydrogenase. *Biochem. J.* **309,** 891–898.
46. McDonald, L. J. and Moss, J. (1993) Stimulation by nitric oxide of an NAD linkage to glyceraldehyde-3-phosphate dehydrogenase. *Proc. Natl. Acad. Sci. USA* **90,** 6238–6241.
47. Chen, R.-W., Saunders, P. A., Wei, H., Li, Z., Seth, P., and Chuang, D.-M. (1999) Involvement of glyceraldehyde-3-phosphate dehydrogenase (GAPDH) and p53 in neuronal apoptosis: evidence that GAPDH is upregulated by p53. *J. Neurosci.* **19,** 9654–9662.
48. Goodman, Y., Bruce, A. J., Cheng, B., and Mattson, M. P. (1996) Estrogens attenuate and corticosterone exacerbates excitotoxicity, oxidative injury, and amyloid β-peptide toxicity in hippocampal neurons. *J. Neurochem.* **66,** 1836–1844.
49. Dubal, D. B., Shugrue, P. J., Wilson, M. E., Merchenthaler, I., and Wise, P. M. (1999) Estradiol modulates bcl-2 in cerebral ischemia: a potential role for estrogen receptors. *J. Neurosci.* **19,** 6385–6393.
50. Joe, I., Zheng, J. B., and Ramirez, V. D. (2000) Specific effects of estrogen/progesterone on glyceraldehyde-3-phosphate dehydrogenase activity: sex differences and estrous cycle-dependent changes. *Soc. Neurosci. Abstr.* **26,** 346.12.

Analytical Approaches for Investigating Apoptosis and Other Biochemical Events in Compartmented Cultures of Sympathetic Neurons

*Barbara Karten, Bronwyn L. MacInnis,
Hubert Eng, Yoko Azumaya, Grace Martin,
Karen Lund, Russell C. Watts, Jean E. Vance,
Dennis E. Vance, and Robert B. Campenot*

1. Introduction

The compartmented culture model utilizing rat sympathetic neurons and sensory neurons is gaining wide use in investigations of neurotrophic factor signaling and axonal transport. Previous chapters have focused on the preparation of compartmented cultures. Here we focus on analytical approaches that we have used with compartmented cultures that can be used to investigate apoptosis, intracellular signaling, and other issues. Techniques used for incubating the cultures with radioisotopes and other agents, harvesting cell extracts from the cultures, using the cultures for histochemistry, and producing cultures from individual transgenic mouse pups are described.

2. Retrograde Signaling

The survival and proper function of many types of neurons is dependent on neurotrophic factors that the neurons obtain from the cells that their axon terminals innervate. It is not established

From: *Neuromethods, Vol. 37: Apoptosis Techniques and Protocols, 2nd Ed.*
Edited by: A. C. LeBlanc @ Humana Press, Inc., Totowa, NJ

how neurotrophic factors such as nerve growth factor (NGF) inter-
acting with receptors on the axon terminals produce retrograde
signals that influence neuronal survival and gene expression in
the cell body. The compartmented culture model permits neurotrophic factors to be applied locally to the distal axons and terminals of cultured neurons, and the effects on the cell bodies can be
locally assayed. Therefore, this system has become a major tool in
the investigation of retrograde signaling. Sympathetic neurons
from newborn rats and sensory neurons from rat embryos have
been utilized *(1,2)*. NGF must be present for sympathetic neurons
to survive in culture *(3)*, and withdrawal of NGF causes the neurons to undergo apoptosis *(4)*. The survival of sympathetic neurons is supported when NGF is provided only to the distal axons
in compartmented cultures *(1,5)*. The prevalent hypothesis for the
mechanism of retrograde signaling is that NGF bound to trkA, its
principal receptor, is endocytosed in the axon terminals, and vesicles containing NGF in their lumens and activated trkA in their membranes are retrogradely transported along axonal microtubules
to the cell body. In the cell body the cytosolic domains of activated
trkA activate signaling pathways that access the nucleus resulting
in the prevention of apoptosis and other effects on gene expression (reviewed in ref. *6*). Some evidence obtained with compartmented cultures has supported the hypothesis of the retrograde
transport of a signaling endosome *(7,8)*, but other evidence suggests that some retrograde signals may be propagated too fast to
be carried by retrograde transport *(9)*. It seems clear that the compartmented culture model will continue to play a major role in
investigations of retrograde signaling and other issues, for example, mechanisms of slow axonal transport *(10)* and axonal production of proteins *(11)* and lipids *(12)*.

The detailed methods for culturing sympathetic neurons and
constructing compartmented cultures have been presented elsewhere *(13,14)*. However, the experimental and analytical techniques
applied to compartmented cultures have not been described except
as methods in research papers. Our purpose here is to describe in
some detail how these cultures are used in experimental approaches
to apoptosis and other biochemical processes occurring in the neurons. The focus will be on those approaches with which we have
direct experience. There is a large area of overlap between compartmented culture approaches to apoptosis and other areas such as
retrograde transport and retrograde signaling. Therefore, we will

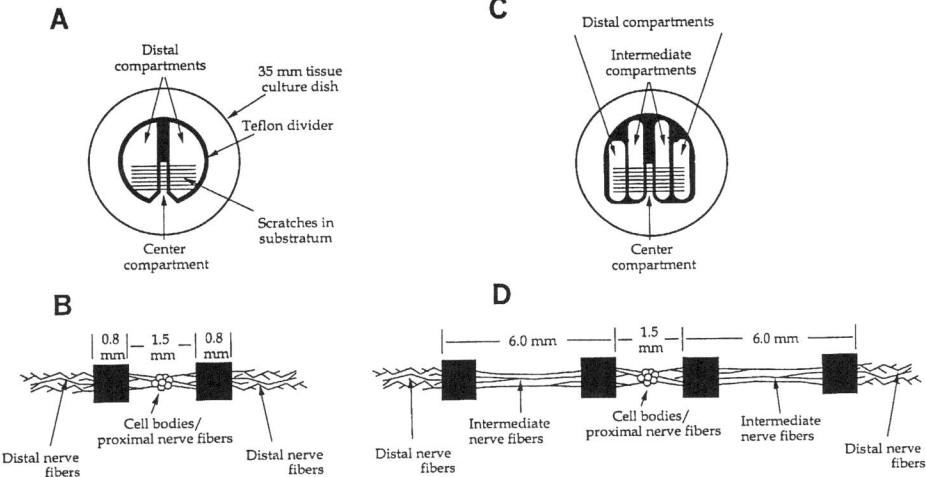

Fig. 1. Schematics of compartmented cultures. (**A**) Three-compartment culture with full-length scratches in the substratum. Scratches can be made to end just inside the distal compartments (see text). The schematic has fewer scratches, which number 21 to delineate 20 collagen strips in actual cultures. (**B**) Enlargement of neurons on a single strip in a three-compartment culture. (**C**) Five-compartment culture. (**D**) Enlargement of neurons on a single strip in a five-compartment culture.

discuss the experimental and analytical techniques in compartmented cultures from a broader perspective.

3. Experimental Analyses Using Compartmented Cultures

3.1. Compartmented Cultures

Compartmented cultures have a unique geometry in which neuronal cell bodies and proximal axons reside in a separate culture compartment from the distal axons. This is achieved by dividing a 35-mm tissue culture dish into several compartments with a Teflon divider seated to the collagen-coated floor of the dish with silicone grease. In the most commonly used three-compartment configuration (Fig. 1), the dish is divided into left- and right-side compartments, with a center compartment between them consisting of a slot that is continuous with the space surrounding the divider. Before assembling the dish and divider, 21 parallel scratches are made in the dish floor with a pin rake after the collagen coating has dried. In this way 20 parallel, collagen-coated tracks are formed between the scratches, which will guide the growing axons to cross under the silicone grease. After making the scratches, a drop of culture medium is placed on the dish surface, located so as to wet the surfaces of the middle regions of the tracks. Then, silicone grease is applied to the bottom surface of the divider through a syringe with an 18-gage needle with a squared-off tip. The divider is then seated onto the dish floor, oriented such that the centers of the tracks run through the center compartment normal to its long dimension and extend into both side compartments. The silicone grease that separates the center compartment and side compartments does not adhere to the wet collagen, while in other regions of the divider the silicone grease adheres firmly to the dry collagen. This results in a gasket-like seal formed by the silicone grease against the wetted surfaces of the tracks. While this seal effectively prevents significant diffusion of medium components from one compartment to another, elongating axons from neurons plated in the center compartments can readily grow beneath the silicone grease and cross into the side compartments.

The methods for obtaining superior cervical ganglion neurons from newborn animals and maintaining them in culture have been described *(15)*. The L15 medium used is thickened with methylcellulose, and cell bodies/proximal axons are provided with medium supplemented with 2.5% rat serum. Distal compartments are provided with L15 medium supplied with 50 ng/mL NGF. Initially center compartments are provided with 20 ng/mL NGF to promote survival and axon growth and with cytosine arabinoside to eliminate non-neuronal cells. After 7 d medium without NGF or cytosine arabinoside is provided to the cell bodies/proximal axons. At this time axons are well established in distal compartments supplied continuously with 50 ng/mL NGF to support axon growth and neuronal survival.

For some experimental designs it is useful to apply treatments to intermediate regions of axons connecting between the cell bodies/proximal axons and the distal axons. Ordinary three-compartment dividers can be used to achieve this by plating the cell bodies in the left compartment and having the axons grow across the center compartment slot and into the right compartment. The method for plating inside compartments has been described *(13,14)*. It is also possible to use five-compartment dividers in which left and right intermediate compartments intervene between the center slot and the left and right distal compartments (Fig. 1).

3.2. Incubations

Most experiments begin with incubating cultures with medium containing experimental agents. If filled, the side compartments each hold 0.5 mL, the center compartment slot can hold 70 µL, and the dish perimeter comfortably holds 1.5 mL. A subtlety that may not be immediately evident is that the center-compartment slot can be filled with medium without also filling the dish perimeter. This is, in fact, how the suspension of neurons is originally plated in the center-compartment slot. Immediately after each compartmented dish is assembled, a small amount of silicone grease is applied to the collagen surface where the mouth of the slot joins the dish perimeter. This produces a hydrophobic barrier on the dish surface such that up to 70 µL of cell suspension can be injected into the center-compartment slot and will not run out into the dish perimeter. The hydrophobicity of the silicone grease at the bottom, and the Teflon on each side, along with the surface tension, causes the column of medium to remain in the slot. After the cells have had sufficient time to settle on the substratum, medium is added to the perimeter of the dish and becomes confluent with the medium in the slot. This allows the medium in the center slot to be conveniently changed simply by aspirating and replacing the medium in the perimeter. Cultures are routinely maintained with medium in the dish perimeter. However, for many experiments this is not desirable. For example, incubations with medium containing radioisotopes such as [^{35}S]methionine to metabolically label proteins synthesized in cell bodies, would be prohibitively expensive and involve excessive radiation levels if each culture were supplied with 1.5 mL of medium at 100–250 µCi/mL. In such experiments we only supply medium in the center compartment slot, just as was done for the original plating of the neurons. The culture medium is aspirated from the perimeter and the center-compartment slot. The tip of the aspirator is made from a sterile, gel-loading micropipet tip. It is helpful to cut off about half of the length of the fine tip to increase the aperture size, since the medium used is usually thickened with methylcellulose and aspiration through an unmodified tip is slow. Any number of ways can be used to attach the tip to an aspirator. It is best to insert the tip at the closed end of the center-compartment slot, away from most of the neurons. Immediately after aspirating each culture, the medium in the center compartment slot is replaced with up to 70 µL of the desired incubation medium delivered with a pipetman fitted with a similarly modified, sterile tip. Incubations of the side compartments can also involve reduced amounts of medium, such as when [^{125}I]NGF is given to the cultures to investigate retrograde transport. Each side compartment is supplied with about 70 μL of medium to ensure coverage of the distal axons. Aspiration and replacement can be accomplished with ordinary micropipet tips or even with Pasteur pipets. To avoid mechanical damage to the axons, care must be taken to flow, not drip, the medium into the compartments.

3.3. Analysis of Cell Extracts

Many experimental protocols involve harvesting of cell extracts once the incubation is completed. If the harvested material is to be used for immunoprecipitation of specific proteins, then the volume is not critical. However, if the harvested material is to be analyzed directly by SDS-PAGE, then it is advantageous to harvest the extracts from multiple cultures in a small enough volume to load in a single gel lane. First, we will discuss the technique of harvesting when the volume is not critical. Cultures are removed from the incubator, placed on a tray of crushed ice, and the medium is removed from all compartments by aspiration as described above. A cut-off gel micropipet tip is used for the center-compartment slot. The medium is replaced with harvesting buffer appropriate to the analysis, e.g., immunoprecipitation buffer. In some experiments the medium bathing the neurons will be collected with a pipet and retained for analysis. After the medium is removed from each culture, the center-compartment slot and the left and right compartments are each given 50 μL of harvesting buffer, which remains in the cultures for the time dictated by the particular protocol (e.g., a very short time in the case of SDS sample buffer; 30 min on ice on a rocker for immunoprecipitation buffer).

Collection of the cell extracts is accomplished by a combination of scraping and trituration. A microfuge tube is set up in crushed ice for each sample to be harvested, for example, one tube for the cell bodies/proximal axons in the center-compartment slots and one tube for the distal axons pooled from the left and right compartments. A cut-off, gel-loading micropipet tip is placed in each centrifuge tube, and a micropipet is set at a volume of 100 μL. The extracts that are most crucial to the experiment are collected first. For example, if the experiment involved [^{35}S]incubation of the cell bodies/proximal axons and anterograde transport of labeled proteins into distal axons, then the distal axons would be harvested first. Conversely, if the experiment involved [^{125}I]NGF incubation of distal axons and retrograde transport of [^{125}I]NGF to cell bodies/proximal axons, then the cell bodies/proximal axons would be harvested first.

To collect left- and right-compartment extracts, the gel-loading micropipet tip from the appropriate centrifuge tube is held with the fingers and the tip is scraped along the floor of the left (or right) compartment for about 15 strokes with a motion parallel to the long axis of the center compartment slot. This process is repeated in the opposite-side compartment. Then the micropipet tip is mounted on the micropipet, and the fluid in the left (or right) compartment is triturated back and forth in such a way as to remove any remaining cellular debris from the entire floor of the compartment. About 5 cycles are sufficient. Then the fluid is collected in the pipet and transferred to the microfuge tube. This process is repeated in the opposite compartment. It is sometimes convenient after scraping to harvest left and right compartments by first transferring the fluid from the right compartment to the left compartment, triturating the left compartment, then pipetting all the fluid into the right compartment, triturating the right compartment, then transferring the fluid to the microfuge tube. In this way the volume triturated is doubled, which can make the procedure easier. Sometimes the micropipet tips become plugged with silicone grease during this procedure. If that happens, it is usually possible to eject any residual fluid from the pipet tip by applying positive pressure with the micropipet set at an increased volume. Once the fluid in the tip has been ejected back into the compartment being harvested or into the centrifuge tube, replace the tip with a fresh one, and continue the harvesting procedure.

To harvest center-compartment slots, the gel-loading micropipet tip from the appropriate microfuge tube is held with the fingers and the tip is scraped across the floor of the center compartment. To avoid losing extract, the floor is scraped in the direction from the mouth of the center slot toward the closed end. About 5 strokes are sufficient. Then the same tip is mounted on the micropipet, the fluid is very carefully triturated two to four times and removed to the appropriate microfuge tube. If the micropipet tip becomes clogged, the same procedure is followed as described in the previous paragraph.

After the culture is completely harvested, it is examined under the inverted microscope to verify that no cell debris remains. Then the next culture is harvested into the same tubes following the same procedure. This procedure is repeated until all the cultures to be pooled have been harvested.

During this procedure, care must be taken not to disturb the divider. If the divider is inadvertently pushed with the harvesting pipet tip, the divider may slide slightly, which destroys the separation between the compartments. If this happens, harvesting must be discontinued and no further extracts from the culture can be added to the tubes without invalidating the results. A fresh micropipet tip is used for the next culture. After the procedure is complete, the number of compartments harvested for each sample is recorded. Volumes can be adjusted later to equalize the culture numbers. For example, if six left/right compartments and five center compartments were successfully harvested for a group, one-sixth of the volume of the left/right extracts can be removed before analysis. With practice, it is possible to harvest cultures reliably without losses.

Sequential harvesting procedures are used when extract volume must be kept low. In such cases, for example, the first culture to be harvested in a group is given SDS sample buffer with 60 μL supplied to the center compartment and each side compartment. Then, the extracts are harvested as above. Instead of fresh SDS sample buffer, the extracts from the first culture are used to harvest the second culture, and so on until all the cultures to be pooled are harvested. There will be some increase in volume of the extracts because residual fluid is present in each culture that is harvested. Therefore, it is helpful to use the starting SDS sample buffer at double strength. A high harvesting proficiency is necessary for sequential harvesting, since disturbing the divider in a culture containing extract will lose the material contributed by all the cultures harvested previously into that extract. With practice sequential harvesting can be carried out routinely.

3.4. Histochemistry

Histochemical analysis of experimental treatments offers several advantages when working with compartmented cultures. Foremost is the cost of analysis: because of the limited amount of material available to be analyzed, pooling of extracts from many cultures is often required to assess proteins by immunoprecipitation and/or

Western blotting techniques. Alternatively, most histochemical analyses can give results on a "per-culture" basis. As with other cell culturing systems, immunostaining or immunofluorescence are convenient and effective ways to assess qualitatively the presence and/or activation state of specific proteins. DAPI or Hoechst DNA stains, which identify DNA condensation and fragmentation, hallmarks of apoptosis, are also useful in experiments when apoptosis is being assayed. Another convenient and useful method for assessing cell viability is the MTT assay. This colorimetric assay is based on the conversion of tetrazolium salt (MTT) to a visible blue formazan product in active mitochondria of live cells. Since mitochondria are present throughout the cell, including the distal axons, evidence of cell death can be visualized in both the axons and or the cell bodies. Since MTT measures the mitochondrial activity of the cell, it serves as an excellent marker for cell survival, but does not distinguish between cell death due to apoptosis vs necrosis.

The techniques involved in processing compartmented cultures for histochemical analysis are, for the most part, similar to those used for other cells in tissue culture. However, special considerations do have to be made when preparing neurons grown in compartments for microscopy. Following experimental treatment cells are fixed with the Teflon divider in place. Both cold acid–alcohol and paraformaldehyde (4%) fixation have been used successfully in our laboratory. After fixation, the divider can be gently removed using a hemostat, although it is often advantageous to keep the divider in place during subsequent incubations. This allows a smaller volume to be used when reagents are limited or expensive, as they can be confined to the region of the neuron that is of interest. Mechanical stress caused by the many washing steps involved in preparing cells for histochemical analysis can cause "lifting off" of the cells from the substratum, even after fixation. Distal axons seem to be especially vulnerable to this effect. To minimize this risk it is necessary to use extreme caution when exchanging the medium bathing the cells for each incubation/washing step. The solutions should be flowed into the compartments, not dripped. Additionally, confining the axons between the scratches in the side compartments seems to make them vulnerable to mechanical detachment. Short scratches that span the center compartment but end a short distance after they emerge into the left and right compartments allow the axons to spread out in the side compartments, producing less cabling of axons and better attachment to the substratum. As well, methyl cellulose can be dissolved into the buffers used in the washing steps, thus increasing the viscosity of the medium and reducing the shear force exerted on the axons.

Once the cultures are ready to be mounted, the divider must be carefully removed with a hemostat. After the divider is removed, any residual medium is aspirated and excess silicone grease is removed from the surface of the dish by wiping with a lint-free tissue wrapped around a pipette tip. Approximately 30–50 μL of mounting medium is sufficient to cover the neurons. Once the neurons are cover slipped, they can be examined through the cover slip with an epifluorescence microscope. It is less favorable to view the neurons through the plastic floor of the dish, which displays some autofluorescence and is too thick for the working distances of most objectives. A potential alternative approach to produce cultures in which living neurons can be viewed from beneath is to cut a hole larger than the diameter of the Teflon divider in each tissue culture dish, attach a glass cover slip over the hole with Silgard (Dow Corning), and coat the inside surface of the glass cover slip with collagen to form the substrate for attachment of neurons. However, it is not possible to pattern the collagen into strips with a pin rake on glass. Some other approach to produce the collagen strips, such as evaporating metal onto collagen surfaces covered with a mask, would have to be used *(16)*.

4. Preparation of Sympathetic Neurons from Superior Cervical Ganglia of Transgenic Mice

Mouse neurons are cultured under essentially identical conditions as rat neurons. If homozygous transgenic mice are viable, then culturing homozygous litters from homozygous breeding pairs is essentially the same as culturing wild-type neurons. However, in cases where homozygous animals are not viable to adulthood, cultures with neurons homozygous for a transgene are produced from the offspring of heterozygous breeding pairs. This is accomplished by plating the neurons isolated from ganglia of each mouse pup into separate cultures. Then the genotype of neurons in each culture is determined by polymerase chain reaction (PCR) analysis of DNA isolated from the tails of the mouse pups. In this way homozygous cultures, heterozygous cultures, and cultures not bearing the transgene are identified. Alternatively, pups could be genotyped from tail samples taken before culturing, and then ganglia from pups of like genotype could be pooled before dissociation and plating. However, we judged this less likely to succeed, since it involves taking the pups from their mother, taking tail samples, marking pups for identification, and then returning them to their mothers to await genotyping. Mothers may not tolerate this manipulation.

To accomplish our procedure, superior cervical ganglia are dissected from newborn pups no more than 2 d old as is routine for all culturing. The ganglia pairs from each pup are placed into separate, sterile microfuge tubes containing 1.2 mL of plating medium (see ref. *15* for medium recipes). Throughout the procedure, separate Pasteur pipets are used for each pair of ganglia to avoid transfer of cells between the preparations. After the dissection of all pups is complete, each pair of ganglia are transferred into separate 15-mL sterile, disposable centrifuge tubes, washed once with 3 mL of phosphate-buffered saline (PBS), and then incubated in 500 μL of 1-mg/mL type 1 collagenase (Sigma) in PBS at 37°C for 25 min. Then 35 μL of a 1% trypsin (Sigma) solution in PBS are added, and the ganglia are incubated for an additional 4 min. A prolonged incubation with trypsin reduces cell viability. The enzyme solution is aspirated carefully with a Pasteur pipet. Ganglia are then washed successively with 3 mL of PBS containing 15% horse serum, twice with 2 mL of plating medium containing 10% horse serum, and once with 2 mL of plating medium without serum. The pair of ganglia from each mouse should be visible throughout the washes. Cells are mechanically dissociated in 0.8 mL of plating medium by trituration with a plugged Pasteur pipet that has been extended over a flame to reduce the tip diameter to approx. 0.5 mm. The cell suspension is centrifuged for 5 min at 800 rpm in an IEC model CL clinical centrifuge. The supernatant is removed carefully, leaving approximately 30 μL in the tube to ensure that the cell pellet is not disturbed. For plating in compartmented dishes, the cell pellet is resuspended in 180 μL of culture medium. Equal volumes of cell suspension (approximately 70 μL each) are plated with a gel loading tip into the center slots of three cultures. The cultures are maintained similarly to ordinary cultures. The genotypes of the cultures are obtained from the tail samples before the cultures are used in experiments.

5. Conclusion

We have presented several detailed procedures that we have found useful in employing compartmented cultures for investigating apoptosis and other areas. The major advantage of compartmented cultures is the ability to locally expose different regions of the neurons to experimental treatments and to analyze the regions separately. This allows local signaling mechanisms in axons and retrograde signaling mechanisms that transmit neurotrophic signals from axon terminals to cell bodies to be investigated. Although the use of primary neurons makes compartmented cultures expensive and time-consuming to produce and small amounts of material must be analyzed, they are used by many laboratories to address issues that are hard or impossible to approach in other ways.

References

1. Campenot, R. B. (1977) Local control of neurite development by nerve growth factor. *Proc. Natl. Acad. Sci. USA* **74**(10), 4516–4519.
2. Sussdorf, W. S. and Campenot, R. B. (1986) Influence of the extracellular potassium environment on neurite growth in sensory neurons, spinal cord neurons and sympathetic neurons. *Brain Res.* **390**(1), 43–52.
3. Chun, L. L. and Patterson, P. H. (1977) Role of nerve growth factor in the development of rat sympathetic neurons in vitro. I. Survival, growth, and differentiation of catecholamine production. *J. Cell Biol.* **75**(3), 705–711.
4. Edwards, S. N. and Tolkovsky, A. M. (1994) Characterization of apoptosis in cultured rat sympathetic neurons after nerve growth factor withdrawal. *J. Cell Biol.* **124**(4), 537–546.
5. Campenot, R. B. (1982) Development of sympathetic neurons in compartmentalized cultures. II. Local control of neurite survival by nerve growth factor. *Dev. Biol.* **93**(1), 13–21.
6. Campenot, R. B. (1994) NGF and the local control of nerve terminal growth. *J. Neurobiol.* **25**(6), 599–611.
7. Watson, F. L., Heerssen, H. M., Moheban, D. B., et al. (1999) Rapid nuclear responses to target-derived neurotrophins require retrograde transport of ligand-receptor complex. *J. Neurosci.* **19**(18), 7889–7900.
8. Riccio, A., Pierchala, B. A., Ciarallo, C. L., and Ginty, D. D. (1997) An NGF-TrkA-mediated retrograde signal to transcription factor CREB in sympathetic neurons. *Science* **277**(5329), 1097–1100.
9. Senger, D. L. and Campenot, R. B. (1997) Rapid retrograde tyrosine phosphorylation of trkA and other proteins in rat sympathetic neurons in compartmented cultures. *J. Cell Biol.* **138**(2), 411–421.
10. Campenot, B., Lund, K., and Senger, D. L. (1996) Delivery of newly synthesized tubulin to rapidly growing distal axons of sympathetic neurons in compartmented cultures. *J. Cell Biol.* **135**(3), 701–709.

11. Eng, H., Lund, K., and Campenot, R. B. (1999) Synthesis of beta-tubulin, actin, and other proteins in axons of sympathetic neurons in compartmented cultures. *J. Neurosci.* **19**(1), 1–9.

12. Posse de Chaves, E., Vance, D. E., Campenot, R. B., and Vance, J. E. (1995) Axonal synthesis of phosphatidylcholine is required for normal axonal growth in rat sympathetic neurons. *J. Cell Biol.* **128**(5), 913–918.

13. Campenot, R. B. and Martin, G. (2001) Construction and use of compart-mented cultures for studies of all biology of neurons, in *Protocols for Neural Cell Culture,* 2nd ed. (Federoff, S. and Richardson, A., eds.), Humana Press, Totowa, NJ, pp. 49–57.

14. Campenot, R. B. (1992) Compartmented culture analysis of nerve growth, in *Cell-Cell Interactions: A Practical Approach* (Stevenson, B., Paul, D., and Gallin, W., eds.), IRL Press, Oxford, UK, pp. 275–298.

15. Hawrot, E. and Patterson, P. H. (1979) Long-term culture of dissociated sympathetic neurons. *Meth. Enzymol.* **28,** 574–584.

16. Letourneau, P. C. (1975) Cell-to-substratum adhesion and guidance of axonal elongation. *Dev. Biol.* **44**(1), 92–101.

Detection and Analysis of Synaptosis

Greg M. Cole and Karen Gylys

1. Introduction

1.1. Alternative Modes of Synaptic Elimination and Loss in Development, Aging, and Disease

1.1.1. Modes of Synaptic Loss (Synaptic Stripping, Resorption, Autophagocytosis).

Essentially three modes of synaptic elimination have been described (1): (1) loss of synapses in synaptically connected neurons following physiological neuron death during development or hormonally driven reorganization; (2) process (generally axonal) retraction and proteolytic degradation of presynaptic elements involving lysosomal upregulation and autophagy; (3) elimination of intact nerve terminals by glial phagocytosis ("synaptic stripping"), notably in the facial nucleus as a retrograde transneuronal effect following axotomy—here microglial cells play a major role (2,3). Similar phagocytosis of intact terminals can be performed by protoplasmic astrocytes as described below.

1.1.2. Synaptic Elimination in Development

Synapses and neurons are overproduced during development, and their selective elimination (and survival) proceeds to sculpt and organize the final pattern of connectivity. Because the number of connections in higher organisms is far too large to be genetically specified in detail, other mechanisms for establishing the final patterns of connectivity must be employed. Synaptic elimination at the neuromuscular junction has been most intensively studied (4). At the neuromuscular junction, activity-dependent release of proteases and their inhibitors, notably thrombin and protease

From: *Neuromethods, Vol. 37: Apoptosis Techniques and Protocols, 2nd Ed.*
Edited by: A. C. LeBlanc @ Humana Press, Inc., Totowa, NJP

nexin I, have been implicated in elimination *(5)*. In the CNS, a careful study of lysosomal upregulation and autophagy in developing visual cortex found that "lysosomal degradation was unlikely to be directly responsible for the majority of synapses" removed during ontogenesis, but apparently represents a secondary mechanism. The authors conclude, "it is tempting to speculate that the majority of synapses are eliminated by a process similar to the synaptic disconnection which has been reported for afferent synapses on spinal motoneurons phagoycytosed during ontogenesis." That is, the majority of synapses eliminated during development are likely removed by phagocytosis of intact terminals (mechanism 3 above). Presumably, this requires the appearance of some recognition signal for phagocytosis, for example, oxidative damage to lipids of phosphatidylserine (PS) exposure to provide recognition motifs for scavenger or PS receptors. While there is scant evidence for oxidative damage as a factor in development, there is abundant evidence for apoptosis, possibly triggered by a loss of trophic support.

Competition for neurotrophic factors acquired at nerve terminals and required for survival has been hypothesized to control developmental neuron death in the CNS. In this process, plasma membrane lysis and an inflammatory response to intracellular antigens is circumvented through "programmed" apoptotic cell death involving neuronal cell bodies—and their distal processes and terminals. Thus, the molecules essential for proapoptotic signaling should be present at nerve terminals. Evidence that they are, and that they may be focally activated in synapses independent of nerve cell loss, will be the focus of the following discussion. If this process of trophic factor withdrawal were to occur at individual synapses, one could hypothesize that a focal activation of proapoptotic signaling might lead to focal production of recognition signals for phagocytosis and intact nerve terminal elimination.

1.1.3. Synaptic Loss in Aging

Synaptic loss in select vulnerable areas has been described in aging in rodents *(6)* and humans *(7,8)*. Although there are a number of papers describing region and strain-dependent losses or the absence of losses, there is very little published information pertaining to mechanisms involved. However, in the aging rat cortex, synapse loss is accompanied by the appearance of atypical synaptic profiles with few or distorted vesicles; these are engulfed intact by protoplasmic astrocytes *(6)*. Both microglia and astrocytes have been implicated in the phagocytosis of degenerating dendrites following axonal reaction in the thalamus *(9)*. Phagocytosis of degenerating terminals by both protoplasmic astrocytes and microglia has also been described after low-level excitotoxic injury *(10)*.

1.1.4. Synaptic Loss in Disease with a Focus on Alzheimer's Disease

The loss of synapses in selected regions in Alzheimer's disease (AD) has been documented by stereological counting at the ultrastructural level *(11,12)*. Frontal cortical *(13)* and hippocampal synaptophysin loss has been correlated with clinical decline *(8,14,15)*. Postsynaptic *(16)* and multiple presynaptic marker loss has been documented in AD hippocampus *(17)*. The most egregious presynaptic pathology is represented by dystrophic presynaptic neurites with upregulated, lysosomally derived organelles suggestive of the type of autophagy associated with axonal retraction (synapse elimination by process 2 above). On the postsynaptic side, dendritic pathology includes Hirano bodies and neurofibrillary tangle-related tau immunoreactive curly fibers. Tangle-bearing cells have reduced synaptophysin message consistent with some relationship between tangles and synapse loss *(18)*. There is also evidence that tangle-bearing neurons have DNA strand breaks, showing upregulation of pro- and antiapoptotic pathways *(19,20)*. The mode or modes of synapse loss in AD remains uncertain. An ultrastructural study of synaptic pathology in biopsies of frontal cortex of AD brain, a region known to have significant synaptic marker loss *(8)*, revealed several types of pathology including dystrophic neurites, distended terminals with swollen vesicles and dense bodies similar to dystrophic neurites, and terminals showing "moderate to severe loss of presynaptic vesicles" *(21)*. While this ultrastructural evidence of vesicle loss in terminals was not quantified, several quantitative biochemical studies have found a preferential loss of a battery of presynaptic vesicle protein markers with relative preservation of synaptic plasma membrane markers *(22,23)*. These results are consistent with the generation of atypical terminals with few or distorted vesicles similar to the mode of phagocytic synapse loss in aging rat brain as described by Adams and discussed under Subheading 1.1.3.

1.2. Definition of Synaptosis

The term *apoptosis* was coined from the Greek words to describe a falling off of leaves on a deciduous tree, to represent a programmed pathway for death of cells while the multicellular organism as a whole survives. A major distinction between apoptotic and necrotic cell death is that the former process preserves plasma and organelle membranes that are removed intact by phagocytosis. In many instances, a neuron, like a tree, may survive a dying back and loss of arbor and synapses. By analogy, we have suggested the term *synaptosis* to describe the falling off or loss of intact synapses via a specialized mode of synaptic loss involving protease (caspase, calpain?) activation, PS externalization and subsequent phagocytosis of intact terminals. This process is hypothesized to take place in ontogenesis, synaptic remodeling, aging, and disease where intact synapses appear to be lost by a similar process involving protease (caspase or calpain?) activation, PS exposure, and phagocytosis of nerve terminals. Rather than laddering of DNA, the biochemical signature of synaptosis would be the generation of proteolytically cleaved proteins in nerve terminals. This is essentially the third mode of loss described under Subheading 1.1.1., while examples of nonsynaptotic nerve terminal loss would be provided by retraction and autophagic resorption or necrotic lysis. A schematic illustration of synaptosis in Alzheimer's disease is given in Fig. 1.

1.3. Recognition Signals for Phagocytosis and the Evolutionary Value of Selective Synaptosis

The surface markers that allow synapses to be recognized and phagocytosed intact are presumably similar to those used to recognize apoptotic, damaged, or senescent cells. The primary modification is surface exposure of PS, which is normally on the inner membrane leaflet but is externalized as a downstream event following caspase activation *(24)*. Phosphatidylserine (PS) exposure during apoptosis is believed to be due to both inhibition of the aminophospholipid translocase that internalizes PS to the inner leaflet and activation of the phospholipid scramblase responsible for lipid exchange between leaflets. The activation of scramblase can be triggered by caspase cleavage of PKC-delta and its subsequent phosphorylation. Whether or not calpain-mediated PKC cleavage and activation can similarly stimulate synaptic scramblase remains unknown. Externalized PS is then recognized by a recently

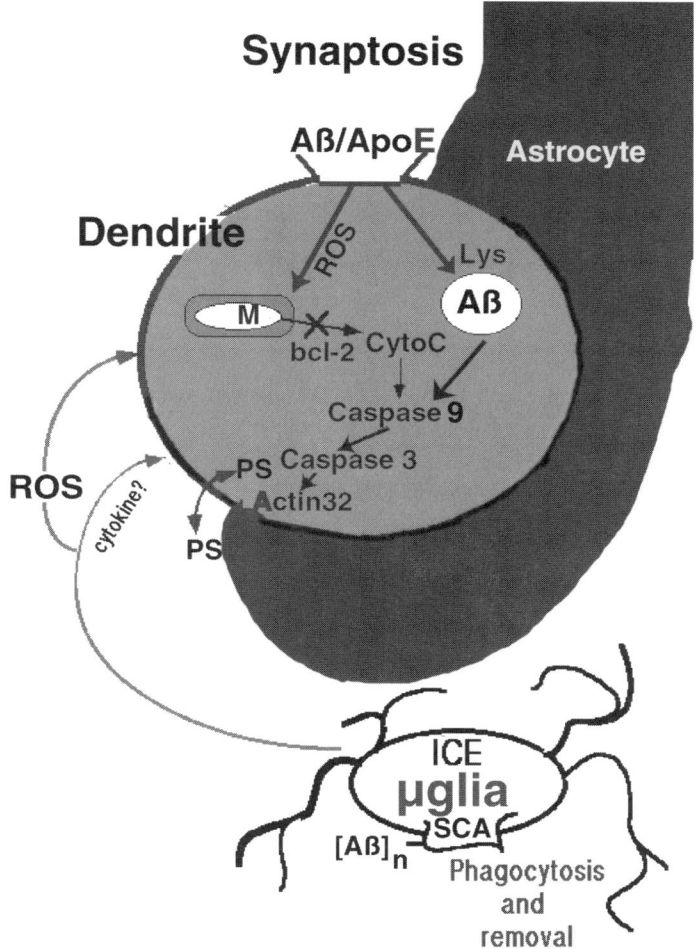

Fig. 1. Synaptosis as a concept is sketched in relationship to an Alzheimer model in which beta-amyloid (Aβ) or its aggregates, [Aβ]n may act to induce focal synaptic damage followed by caspase (or calpain) activation and cleavage of cytoskeletal substrates, phosphatidylserine (PS) externalization, and phagocytosis by astrocytes or microglia (μglia). Several possible mechanisms for Aβ-induced damage are indicated. One candidate pathway involves lipoprotein receptor-mediated uptake into the endosomal lysosomal (lys) system leading to Aβ aggregation, cathepsin D release, and caspase activation. Other pathways involve interaction with scavenger receptors (SCA) on microglia or an undefined Aβ receptor on neurons, leading to release of reactive oxygen species (ROS), oxidative damage to mitochondria (M), and release of cytochrome-*c* (cyto C), followed by caspase activation.

cloned receptor identified by V. Fadok (25) or by scavenger receptors known to recognize oxidatively modified membranes and lipoproteins (26).

1.4. Light Level and Ultrastructural Appearance of Synaptosis

As described below, we generated an antibody to a neo-epitope present on a fragment of caspase-cleaved actin ("fractin") and demonstrated its generation in punctate profiles during apoptosis of cultured differentiated neuroblastoma and on neurons and their processes in AD brain (27). At the light level this antibody labeled swollen and fragmenting neuronal processes, suggesting that "focal and local activation of these proteases (caspases) in neuronal processes...could participate in the trimming and removal of processes by a distal activation of caspase activity." In order to test this hypothesis in the absence of postmortem alterations, we used confocal microscopy to examine aged APPsw transgenic mice and found punctate peri-plaque neuropil staining that co-localized with synaptic markers. This punctate labeling also appeared frequently to overlap activated microglia surrounding the plaques (28), consistent with microglial phagocytosis of dystrophic neurites. Examination of the ultrastructure of plaques in aged APPsw mice revealed microglia apparently engulfing dystrophic neurites (Fig. 2). Similar pictures of phagocytic peri-plaque microglia engulfing dystrophic neurites have been produced using well-preserved aged monkey brain and conventional transmission electron microscopy (29). With our APPsw transgenics, these processes were only lightly labeled with fractin. In collaboration with Antoine Triller and Phillipe Rostaing, we found the most frequent fractin labeling in APPsw transgenics was often synaptic, but largely postsynaptic (Triller et al., unpublished). These images show a progression of postsynaptic fractin staining apparently followed by phagocytosis of intact synapses by protoplasmic astrocytes strongly resembling the type of synapse loss previously described in aging rat brain by Adams and Jones (6). Examples of this punctate synaptic fractin staining in CA1 and CA3 hippocampus of aged APPsw transgenic mice are illustrated in Fig. 3. Synapse and neuron loss have been reported in these vulnerable fields of mutant APP transgenic mice (22). At the ultrastructural level, immunogold labeling identified dendrites and spines as the principal site of labeling (Fig. 4.).

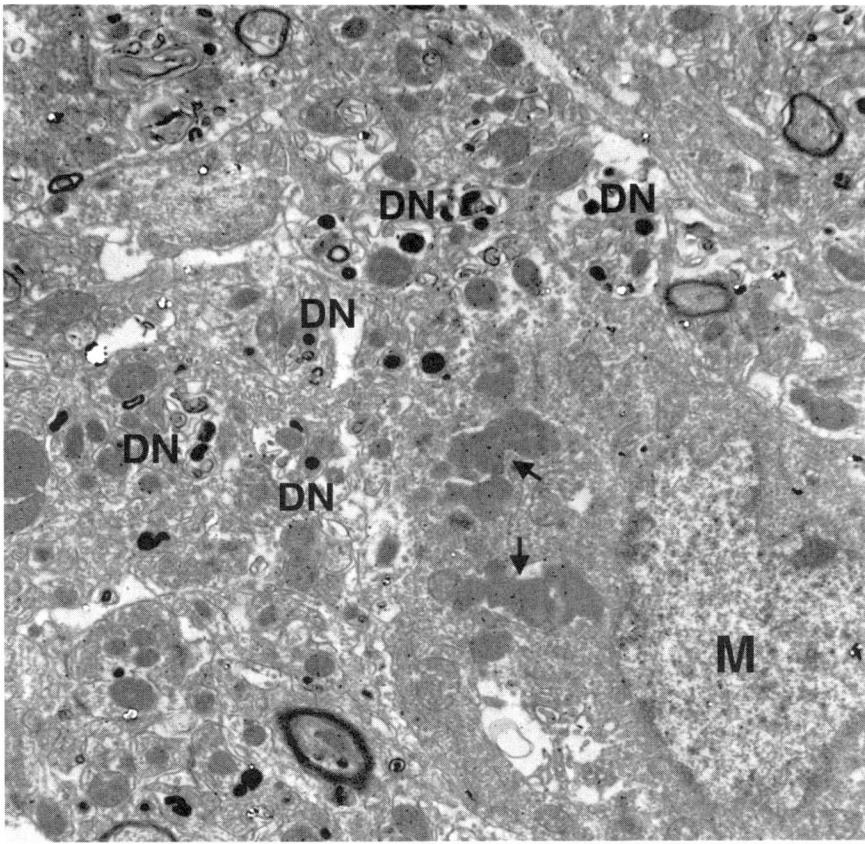

Tg2576 Plaque fractin

500 nm

Fig. 2. Preembedding immunogold labeling for the fractin neoepitope on caspase-cleaved actin in the vicinity of a plaque in an ultrathin section from an aged APPsw (line Tg2576) mouse brain. Dystrophic neurites (DN) show sparse gold labeling. A microglia cell (M) with processes juxtaposed to DN appears to be accumulating fractin label in secondary lysosomes (arrows) consistent with phagocytosis of dystrophic neurites.

2. Methods for *In Situ* Detection of Caspase Activation

Several methods have been employed for the *in situ* detection of caspase activation that could be used to detect apoptosis in synapses and neurites.

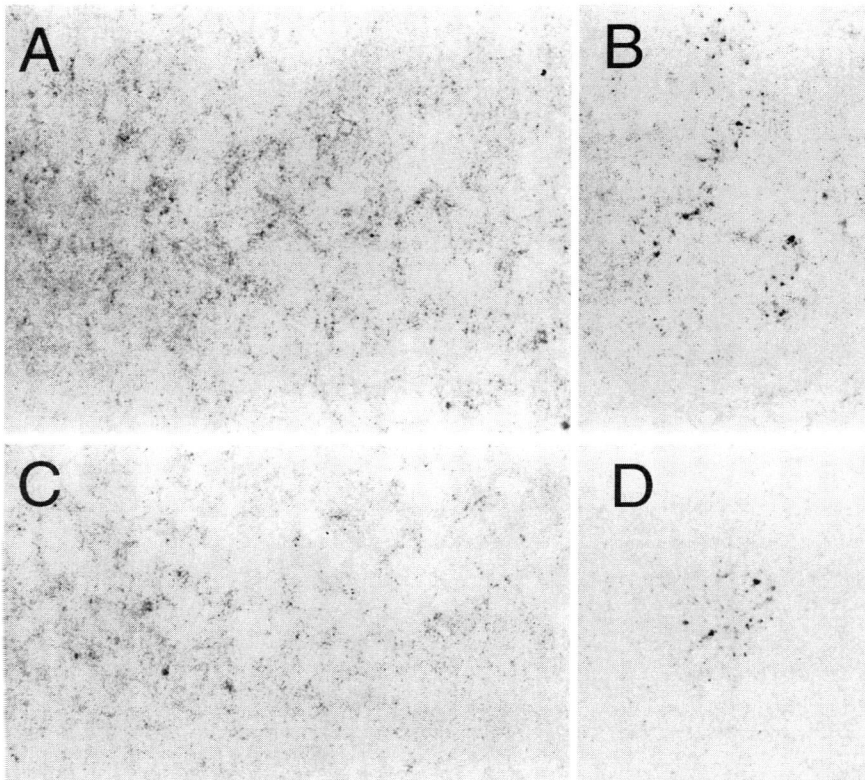

Fig. 3. An example of subtle punctate labeling for the fractin neoepitope on caspase-cleaved actin was found in a somato-dendritic labeling pattern using diaminobenzidine immunocytochemistry in neuronal
layers CA3 (**A**) and CA1 (**C**) of the hippocampus of an aged APPsw mouse.
Similar punctate labeling was found in or around plaques (**B** and **D**) in
both hippocampus and cortex. Orig. mag. = 1320×.

1. Although caspase cleavage of fluorogenic substrates is typically used to assay caspase activation in extracts, the use of such
   substrates for histochemical detection of *in situ* caspase activation has been limited. Affinity labeling of activated caspases
   with biotinylated active site-directed inhibitors has been used
   by several groups for *in situ* detection of activated caspases
   (*30–32*). An example of the method applied to dendrites and
   synapses is briefly outlined under Subheading 4.1.1.

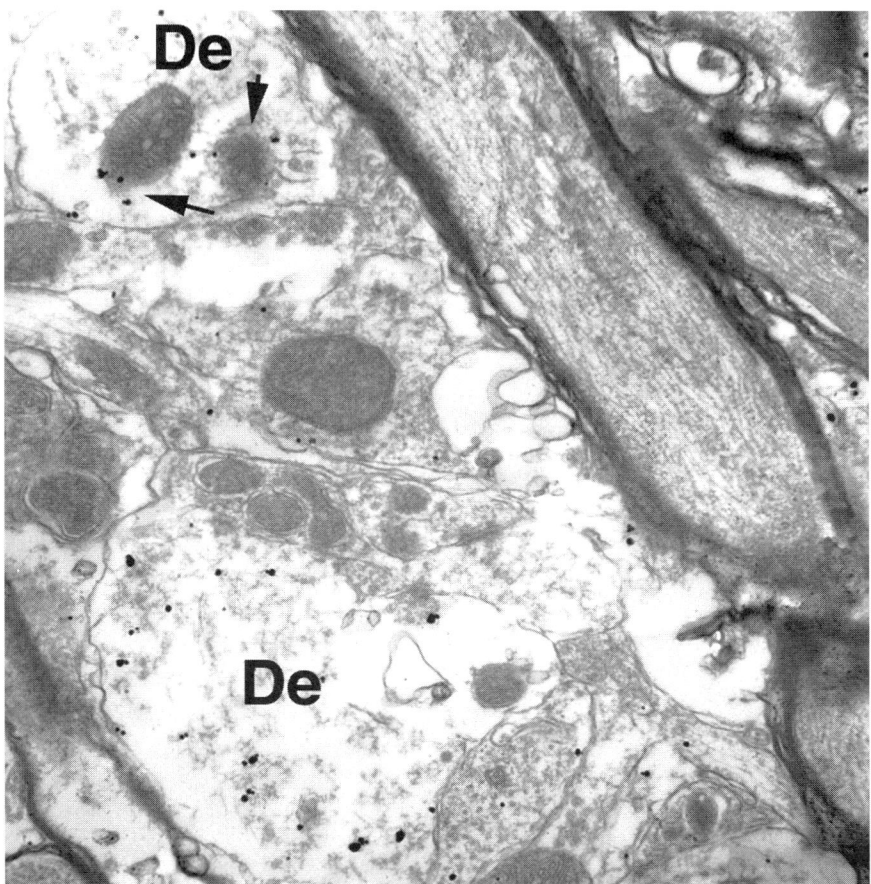

Tg2576 hippo fractin   **500 nm**

Fig. 4. Punctate hippocampal CA labeling of aged APPsw mice with the fractin antibody illustrated in Fig. 3 was determined to be in dendrites by our collaborators P. Rostaing and A. Triller. Here we illustrate a confirmation of this observation obtained by F. Yang in our laboratory using preembedding immunogold labeling for fractin.

2. End-specific antibodies to caspase cleavage products have been developed (27) and applied. In principal, one could use end-specific antibodies to caspase-cleaved substrates found in neurites and synapses to detect focal distal caspase activation. Identified caspase substrates found in synapses include actin,

fodrin, procaspase 9, amyloid precursor protein, presenilin, tau, and presumably many others.

3. Confocal double labeling for synaptic markers. Although it is possible to detect caspase activation at synaptic sites by double-labeling for synaptic markers and antibodies to activated caspases or neo-epitopes present on caspase-cleaved substrates (28), the resolution of confocal microscopy (~0.2 µmol) typically cannot routinely distinguish between closely apposed pre- and postsynaptic sites. Therefore, immunoelectron microscopic approaches are required.

4. Immunoultrastructural methods. We are aware of no published reports on the ultrastructural identification of sites of caspase activation. However, we have used preembedding immunogold labeling to detect caspase-cleaved actin fragment in *APPsw* transgenics, and A. Triller and P. Rostaing have used both immunogold and diaminobenzidine to detect the same actin neo-epitope in primarily postsynaptic sites in aged *APPsw* animals (Triller et al., unpublished observations). As is typical with immunoultrastructural studies, a major difficulty lies in finding sufficiently rapid and strong glutaraldehyde perfusion fixation protocols that adequately preserve ultrastructure while allowing antibody and label penetration. It is essential to use antibodies directed at the neo-epitope containing the free C terminus of cleaved fragments rather than the N-terminal neo-epitope, which will be destroyed by strong aldehyde fixation.

3. Evidence for In-Vivo Caspase Activation in Nerve Terminals in AD

3.1. Fractin Staining (in Ischemia, in AD Brain, and in an Animal Model for AD via Confocal Microscopy and ImmunoEM)

Our group and Siman's group developed the first end-specific antibodies to caspase substrates (33,34) and applied these to CNS tissue (27,35). Srinivasan and colleagues developed the first end-specific antibody to an activated caspase, caspase-3 (36). Since then, a number of other groups have developed end-specific antibodies to activated caspases or their substrates and commercial antibodies have been developed. Most of these antibodies strongly overlapped conventional TUNEL labeling in perikarya of apoptotic neurons in animal models. These end-stage events are in cascasdes involving multiple amplifications and frequently upregulation of caspase expression; they are readily detected with relatively insensitive probes.

As discussed above, a major objective in choosing a caspase-cleaved actin fragment as a substrate was to detect subtle initiating caspase activation in neuritic processes and synapses. This requires detection of small amounts of activated preformed caspase or products without amplification from cascade-driven increased expression in dying cells. For that goal, we chose actin because it is not only the most prevalent and ubiquitous protein in cells, but also a component of bundled cytoskeletal filaments, which should result in more readily discriminated focal concentrations of antigen than diffuse cytosolic substrates. The fractin epitope can also be generated by multiple-effector caspases definitely including 1 and 3 and possibly including 6 and 7 *(37)*. Other caspases may also cleave actin at the ELPD fractin site because fractin-generating caspases 1, 3, and 6 represent each of the three groups of caspases defined by preferred substrate specificities (group I, caspases-1, 4, 5 with substrates WEHD or YVAD; group II, caspases-2, 3, 7, and 10 with substrates DEXD; and group III, caspases-6, 8, and 9 [I,V,L]EXD). If multiple caspases in a cascade can generate the epitope, greater sensitivity will be obtained. Finally, with an abundant substrate, reasonably stable product molecules will likely outnumber the catalyst by orders of magnitude. For all of these reasons, detection of the fractin epitope would be expected to be a relatively sensitive method compared with detection of a specific activated enzyme.

Fractin antibody detects apoptotic events in a number of different systems *(38)*. As discussed above, fractin antibody detected punctate peri-plaque labeling in AD and APP transgenic mouse brain where immunoultrastructural studies identified the most common staining as postsynaptic and dendritic. This type of postsynaptic labeling was also observed in entorhinal cortex and hippocampal CA1 and CA3. Consistent with this, fractin also labels Hirano body inclusions in dendrites in AD brain *(39)*. Recently, AMPA receptor subunits have also been identified as substrates for caspases *(40)*. Intriguingly, abnormal AMPA receptor immunoreactive inclusions have also been identified in dystrophic dendrites in AD brain *(41)*. These results would be consistent with an excitotoxic component to neuritic degeneration, synapse loss, and later neuron loss in vulnerable CA1 and CA3 in AD and at least some APP transgenics *(42)*. However, at 16 mo of age in the Tg2576 *APPsw* transgenics in which we observe CA3 and CA1 fractin labeling, no quantifiable neuron loss in CA1 could be detected *(43)*. These results would be consistent with synaptosis preceding neuron loss in these regions.

In contrast, a light microscopic study with CM1 antibody to activated caspase-3 found no peri-plaque or apoptotic neuron labeling in AD brain or transgenic mouse models (Roth, K.). The reason for these apparently conflicting results may lie in the sensitivity issue discussed above, the generation of fractin by caspases other than caspase-3, or both. For example, unlike caspase-3, both procaspase-6 and a fragment apparently representing activated caspase-6 are found in postmortem AD brain and caspase-6 activation and APP cleavage occur in cultured human neurons prior to commitment to apoptosis; caspase-6 cleaves actin to generate a 32-kDa fractin-like band *(37)*.

3.2. Other Examples of Punctate
Peri-Plaque Caspase Cleaved Substrate Labeling

Several synaptic caspase substrates have been identified including amyloid precursor protein (APP) *(37)* and presenilin 1 *(44,45)*. Using an end-specific antibody to a caspase-cleaved APP fragment, Gervais et al. *(46)* demonstrated punctate peri-plaque labeling in AD brain. Although consistent with neuritic labeling, no attempt to identify the profiles was made.

3.3. Analysis of Activated Caspase
in Synaptosome Preparations from AD and Control Brain

Amyloid precursor protein cleavage by activated caspases generates a C-terminal peptide product that was shown to be toxic to cultured neurons. Although attempts to detect activated caspases in extracts from AD brain homogenates failed to reveal detectable increases in activated caspase on Westerns, activated caspase-9 was clearly detectable and elevated in synaptosome fractions from AD brain compared with controls using Western analysis *(47)*. This result provides direct support for significant activation of initiating caspases in the cascade in nerve terminals in the absence of a major upregulation of perikaryal caspases typical of apoptotic cells.

4. Analysis of Synaptosis In Vitro

4.1. Cell Culture Approaches

4.1.1. Primary Neuronal Cultures and Cell Lines

Some of the initial evidence that focal neuritic and synaptic activation of caspases can occur came from studies with cultured hippocampal neurons carried out by Mattson's group. These investigators exposed primary hippocampal neurons matured for 7–9 d to toxic concentrations of staurosporine or glutamate *(48)* or β-amyloid aggregates *(49)*. The cells were affinity labeled for activated caspase. They were exposed to 0.02% digitonin in the presence of biotinylated DEVD and then fixed, permeabilized with Triton X-100, and stained with fluorescently tagged streptavidin. This treatment revealed related increases in dendritic and cell body-activated caspase interpreted as caspase-3 based on DEVD specificity. Dendritic biotin-DEVD labeling preceeded cell body labeling and focal application of glutamate or Aβ "with a puffer pipet" resulted in increased focal dendritic labeling.

4.1.2. Campenot Chambers

Specialized culture techniques have been developed in which the neurites (viz., axons) of sympathetic ganglia maintained with NGF are separated from the cell bodies by a sealed barrier *(50)*. These chambered cultures have been used to demonstrate focal effects of NGF application to the neurite compartment without exposure of the cell bodies *(51)*. Using this system, Ivins, Cotman, and collaborators applied toxic β-amyloid peptide aggregates exclusively to neurites to provoke focal neuritic degeneration *(52)*. Annexin V binding detected extracellular PS exposure that was blocked by zVAD-fmk, and caspase activation was detected in the neurites exposed to β-amyloid. In contrast, Raff's group used Campenot chambers to demonstrate that although caspase activation occurred in cell bodies and axons after NGF withdrawal, focal withdrawal of NGF from compartmentalized dorsal root ganglia axons did not result in caspase activation *(53)*.

4.1.3. Organotypic Hippocampal Slice Cultures

Organotypic hippocampal slice cultures (OHSC) have many advantages for study of phenomena related to neurodegeneration and aging *(54–58)*. Methods for their use have recently been reviewed

(59). We have found extensive fractin labeling in cell bodies, and neuronal processes occurs in the first 3 d following slice preparation (unpublished results). TUNEL-positive cell body labeling co-localizes with fractin labeling. While there was reduced labeling after 1 or 2 wk in vitro, in the cultures examined the background labeling remained too high for use as a satisfactory tool for investigating experimentally induced loss of synapses and arbor using the relatively sensitive fractin probe.

4.2. Synaptosome Preparations

4.2.1. Gradient Purified Synaptosome Preparations

Homogenate preparations of nerve terminals have proven a useful model system for the study of cellular events in neurodegeneration, in particular for the study of mechanisms related to the production of reactive oxygen species by amyloid-β protein *(60)*. In addition to the primary culture system described under Subheading 4.1.1., Mattson and colleagues have used purified synaptosomes to demonstrate caspase activation and increased annexin labeling following treatment with staurosporine, glutamate, and β-amyloid *(48,49)*. In these papers, the authors also measured altered mitochondrial function in synaptosomes following apoptotic treatments using probes for mitochondrial membrane potential and mitochondrial ROS levels in isolated nerve terminals.

4.2.2. Flow Cytometry of Intact Synaptosomes

Historically, a primary disadvantage of synaptosomal preparations has been the inability to distinguish measured outcomes in intact synaptosomes from those occurring in contaminants of the preparation (free mitochondria, myelin, membrane fragments, etc). Additionally, the identification of neuronal vs glial populations and of neurochemical subpopulations, for example, cholinergic terminals, has not been possible. The use of flow cytometry techniques (fluorescence-activated cell sorting, or FACS) for quantitative study of homogenate preparations of nerve terminals was first described by Wolf and colleagues in 1989. This group characterized the purified synaptosomal preparation and used this technique for the study of dopaminergic terminals in subsequent papers *(61–64)*. The advantages of flow cytometry analysis of terminals include the rapid analysis of large populations, double and triple labeling designs, and quantitative measure of changes in brightness of fluorescence in objects that are too small to assess readily by light microscopy.

4.2.2.1. P2 Preparations and Clean-up by Gating

The crude synaptosomal preparation, or P-2 pellet, offers the advantages of faster preparation time, greater yield, and increased functional integrity. Therefore we have developed procedures for flow cytometric analysis of crude synaptosomes (65). An antibody directed against the presynaptic docking protein SNAP-25 labeled 66% of the particles in this fraction, indicating neuronal origin. The glial marker GFAP labeled 35% of particles, and 90% of the particles were shown to be intact and esterase-positive using the viability dye calcein AM. Consistent with electron microscopy studies of synaptosomal preparations (66), the presence of both free and intrasynaptosomal mitochondria is indicated by the labeling of 74% of particles with nonyl acridine orange.

When an analysis gate is drawn to include only the largest particles (greater than 1 μm based on size standards), most contaminants are excluded and quantitative data are obtained from a population that is both (1) greater than 95% neuronal (SNAP-25-positive) and (2) virtually 100% intact (calcein AM-positive). In this way, "purification" of intact synaptosomes is accomplished during data analysis (Fig. 5). This level of purity in the analyzed material is better than can be achieved by density gradient purification methods. One may also use a combination of a synaptosomal marker and size to achieve essentially 100% purity.

4.2.2.2. Rapid Kinetics of Calcein AM
and Annexin Double Labeling of Intact Terminals

In order to study markers of apoptosis in the intact neuronal population of interest within a synaptosomal homogenate, we have developed a novel dual-color assay that is analogous to standard flow cytometry assays for detection of apoptosis in intact cells (67). In our method the viability dye is calcein AM, which stains only intact particles with esterase activity; "apoptotic" particles are positive for both calcein AM and annexin-V (Fig. 6), which binds to externalized PS. The number of double-labeled particles was increased after a 1 min incubation, and continued to increase for the duration of the 3-h experiment; after 3 h, approximately 50% of the intact synaptosomes were annexin-labeled. Note that annexin-V binding is consistent with, but does not necessarily

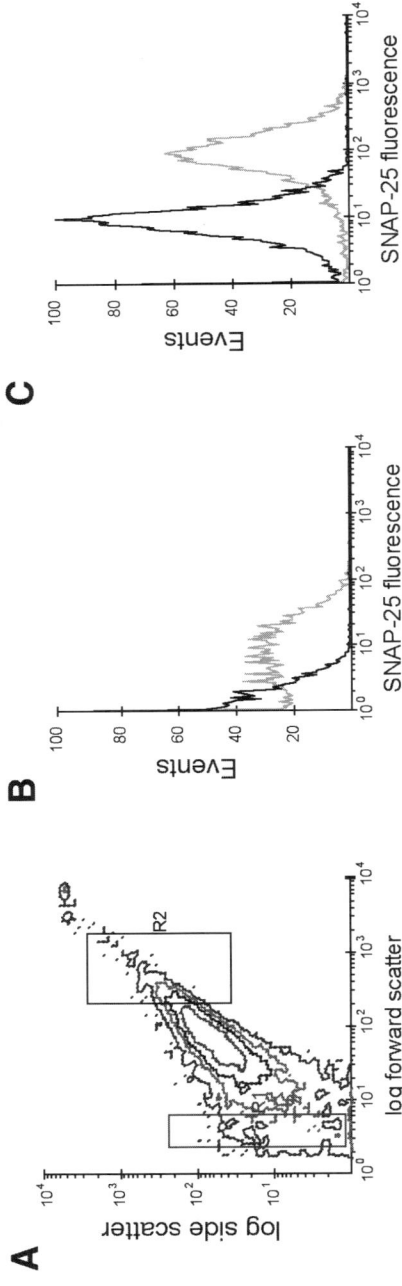

Fig. 5. (A) Flow cytometric analysis of forward and side scatter of a rat P2 synaptosome preparation is used to selectively "gate" out the small debris and focus the analysis on larger 1–4-μmol particles (box R2) with significant internal vesicle populations reflected in side scatter. Thus, as shown (B) by SNAP-25 immunofluorescence (light line), many of the smaller particles (box R1) are negative for this synaptic plasma membrane marker (overlapping the dark line showing nonspecific IgG control fluorescence). In contrast, 95% of the particles in box R2 (C) are SNAP-25 positive. This "clean-up" in the gating allows analysis of events in synaptosomes without density gradient purification.

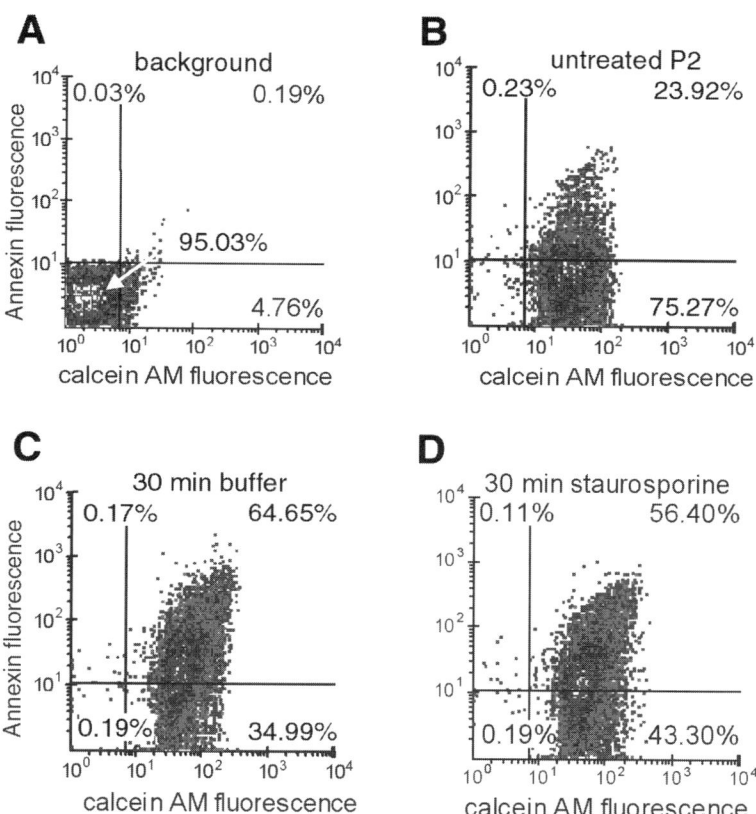

Fig. 6. Scattergrams illustrate results of FACS analysis of cortical calcein AM (a viability marker for intact membranes with fluoroscein optics) and annexin-V rhodamine fluorescence in rat P2 fraction synaptosomes with and without staurosporine treatment. (**A**) Settings for background fluorescence of synaptosomes gated on R2 in the absence of reagents. (**B**) Shows that 99.77% of freshly prepared P2 fraction gated on R2 are positive for calcein AM fluorescence, which requires intact membranes. Surprisingly, nearly 24% of this prep shows annexin-V labeling consistent with phosphatidylserine externalization on intact terminals. (**C**) Shows a 2.7-fold increase in annexin-labeled, intact terminals after a 30-min incubation in buffer. In a 30-min incubation with staurosporine (STS) at doses sufficient to induce apoptosis in SY5Y neuroblastoma (not shown), buffer alone increased the number of annexin V-positive terminals to the same degree as STS (**D**).

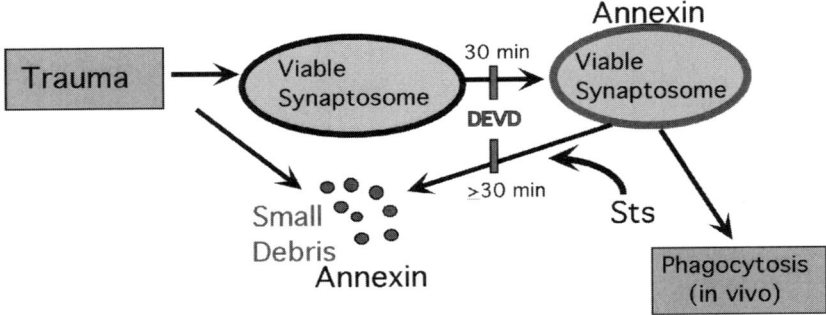

Fig. 7. Summary diagram of flow cytometry of synaptosis. The trauma involved in the synaptosome and debris preparation induces a rapid increase in annexin-V labeling of viable terminals, which peaks at about 30 min. The rapid increase precludes density gradient purification steps. Treatment with DEVD peptide-based caspase inhibitors can reduce but not prevent this annexin-V labeling. Incubations longer than 30 min result in significant losses of large viable particles to the annexin positive debris fraction. Staurosporine (STS) promotes this loss of viability, while DEVD caspase inhibitors only partially protect against this loss of large, annexin-positive, intact terminals, which would normally undergo rapid phagocytosis in vivo.

imply, prior caspase activation as calpain and perhaps other proteases may achieve the same endpoint as discussed later. Interestingly, staurosporine treatment produced a larger increase than buffer only in short incubations (less than 30 min). This suggests that an early-feed-forward activation of protease pathways leading to PS externalization occurs during tissue homogenization. Such a self-amplifying process is consistent with the idea that apoptotic proteolytic cascades are a nonlinear process exhibiting on–off switching rather than gradual activation *(81)*. The speed observed for the loss of membrane asymmetry is an indication of the potential in neurites for a rapid response to toxic insults. A schematic diagram illustrating the detection of annexin V labeling of intact terminals by flow cytometry is given in Fig. 7.

4.2.2.3. Caspase Activation

Rapid caspase activation underlying the annexin-V increase described above is suggested by our finding that increases are observed in caspase-1 activity and caspase-3-like activity in the crude synaptosomal preparation *(67)*. Moreover, peptide inhibitors of caspase activity (Ac-YVAD-CHO and AC-DEVD-CHO) resulted in a partial blockade of increased annexin-V fluorescence in a 30-min incubation (Gylys et al., unpublished observations). However, we were unable to detect a consistent increase in fractin labeling in these predominantly presynaptic nerve terminal preparations.

4.2.2.4. ROLE OF CASPASE ACTIVATION
      IN SYNAPTOSOME DEGENERATION

Flow cytometry analysis demonstrates that larger synaptosomes become lysed during longer incubations. Our recent experiments (Gylys et al., unpublished observations) have shown that this degradation can be blocked by addition of the broad-spectrum caspase inhibitor z-VAD-fmk (300 μ*M*). Because z-VAD-fmk inhibits calpain as well as caspases *(68)*, this result, considered in light of the modest protection of annexin-V staining by caspase inhibitors described above, may reflect the activation of multiple proteases in terminals exposed to apoptotic insults.

4.2.3. Potential for Analysis of Postmortem Human Synaptosomes

Synaptosomes may be prepared from cryopreserved postmortem human tissue *(66)*, and this preparation has been used to demonstrate amyloid-B toxicity to ion-motive ATPases *(69)*. Experiments in our laboratory show that approximately 70% of human synaptosomes are viable, e.g., calcein-AM positive (unpublished observations). This compares to 90% integrity observed in freshly prepared synaptosomes from rat brain *(65)*. Preliminary flow cytometry studies indicate that a postsynaptic marker, PSD-95, is decreased in AD brain compared to control brain, and that antibodies directed against β-amyloid show increased labeling in nerve terminals from Alzheimer's brain, suggesting that analysis of human synaptosomes may provide unique insights into AD-related mechanisms of terminal degeneration.

5. Possible Mechanisms Leading to Synaptosis

5.1. Focal Trophic Factor Withdrawal

It is reasonable to hypothesize that because total neurotrophic factor withdrawal leads to global caspase activation and apoptotic death in many neural systems, focal trophic factor withdrawal might lead to focal synaptic and axonal caspase activation. This is consistent with a number of observations cited above. However, caspase activation has not been observed in sciatic and optic nerve or retinal ganglion cells undergoing Wallerian degeneration after axotomy or after focal withdrawal of NGF from DRG axons as discussed under Subheading 4.1.2. *(53).* In contrast, NGF withdrawal from the cell body resulted in caspase-3 activation in both the cell body and processes, suggesting that procaspases are present in processes, but remain inactive. Wallerian degeneration is accompanied by sustained elevations of free calcium and calpain activation. As shown in models for excitotoxicity, calpain activation can block the activation of caspases, notably caspase-3, by cytochrome-*c (70).* Calpain cleaves caspase-3 (and calpastatin), and these effects on upstream apoptotic effectors may play a role in prevention of caspase activation. This preventive effect stops caspase activation despite cytochrome-*c* release downstream from trophic factor withdrawal. These events may account for the lack of caspase activation in some of the preparations *(53).* Whatever the explanation, the lack of caspase activation in axons after trophic factor withdrawal may still allow focal selective synaptosis to occur. This effect could involve focal synaptic caspase activation, but is more likely mediated by calpain (see below). The absence of axonal caspase activation in other models is consistent with our in-vivo observations of selective caspase activation indicated by fractin labeling on the postsynaptic, somato-dendritic side.

5.2.1. Excitotoxicity, Persistent Elevations of Calcium, and Calpain Activation

As discussed under Subheading 4.1.1., there is evidence that glutamate can induce focal dendritic caspase activation in tissue culture and that this can be inhibited by z-VAD-fmk *(48).* This inhibitor is similarly active against caspases and calpain *(68)* In nucleus-free systems such as platelets, elevated calcium recapitulates apoptotic events, including cell shrinkage, plasma membrane microvesiculation, PS externalization, and proteolysis of procaspase-9, procaspase-3, gelsolin, and protein kinase C-δ *(68).* However, these events occur without caspase activation, and most can be inhibited by calpain inhibitors. In thymocytes and neutrophils, calpains have been implicated as essential mediators of apoptosis *(71,72).* In fact, there are substantial parallels between calpain and caspases as twin effector proteases in apoptotic signaling in a variety of cells, including neurons *(73).* These include major overlaps in substrates including spectrins, actin, tau, vimentin, CaM kinases, focal adhesion kinase, PARP, and the endogenous calpain inhibitor, calpastatin. Caspase-3 cleavage of calpain inhibitor, calpastatin, and calpain cleavage of casapse-3 are channels for crosstalk that regulate both proteases in many apoptotic preparations. Both calpain and caspase inhibitors can offer significant neuroprotection *(74,75)*. Thus, either or both caspase and calpain activation may be effectors of synaptosis in different systems. For example, presynaptic terminal loss may be primarily effected by calpain activation leading to membrano-cytoskeletal breakdown, PS exposure, and phagocytosis, while similar events on the postsynaptic side may depend more on caspase activation or both proteases. Evidence for spectrin breakdown to the 150-kDa calpain cleavage product has been observed in AD *(76)* and correlated with abnormal neuritic spectrin labeling and synaptophysin loss. More recently, calpain activation in neurotoxicity has been linked to p35-to-p25 conversion and activation of a CdK5 tau kinase in AD brain *(77)*. Other evidence for a link between elevated calpain activation in AD and tangle formation has been recently reviewed *(78)*.

5.2.2. Oxidative Damage

One probable mechanism promoting caspase activity in neurodegenerative conditions is oxidative damage to the mitochondria leading to cytochrome-*c* release and other events initiating caspase activation. Oxidative damage is implicated in ischemia/reperfusion, Parkinson's disease, Alzheimer's disease, and aging *(79)*. In AD, oxidative damage may play a critical role *(80)*. If this damage is initiated in synaptic compartments, it could lead to synaptosis. This would be consistent with mitochondria as both the primary source of free radical generation and initiation of apoptosis *(81)*. Free iron, a potent pro-oxidant molecule, appeared to stimulate apoptotic events in synaptosomes *(48)*. Similarly, like free iron, Aβ induces both oxidative damage to mitochondria in synaptosomes *(82)* and apoptotic events *(49)*.

5.2.3. Aβ Toxicity and Synapse Loss

There are at least four ways β-amyloid peptide could lead to synaptosis and synapse loss in AD brain. Since Aβ is generated at synapses, these events may take place in either the pre- or postsynaptic compartments, resulting in synaptosis. First, direct aggregated Aβ toxicity promotes oxidative damage in some systems *(83)*, including nerve terminals *(84)*. Second, Aβ can also directly activate comple-

ment leading to attack complex formation and calcium influxes
(85,86). Third, β-amyloid internalized may accumulate in the endo-
somal-lysosomal system *(87)* and potentially lead to lysosomal leak-
age of cathepsin D, a protease able to activate a caspase cascade
(88,89). Finally, soluble Aβ (possibly oligomers or ADDLs) may act
on NMDA receptor pathways and promote excitotoxic events *(58)*.

6. Summary and Conclusions

The process of glial phagocytosis and removal of intact nerve
terminals is suggested to reflect the focal synaptic activation of proteolytic steps involved in apoptosis leading to cytoskeletal breakdown, scramblase activation, PS exposure recognition by glial
receptors, and phagocytosis. Evidence that this process can proceed involving focal caspase activation in Alzheimer's disease and
amyloid-precursor transgenic mice is discussed in relationship to
synapse loss. Although caspase activation is the signature apoptotic pathway, both caspase and calpain can cleave an overlapping
set of synaptic substrates and the two pathways appear closely
interrelated. The proteolytic processing required to trigger PS exposure and nerve terminal removal may also involve calpain activation. Evidence is accumulating that focal synaptic and neuritic
damage secondary to excitotoxicty, oxidative damage, or β-amyloid accumulation may serve as triggers for synaptosis and nerve
terminal removal in neurodegenerative diseases.

References

1. Wolff, J. R. and Missler, M. (1993) Synaptic remodelling and elimination as integral processes of synaptogenesis. *APMIS Suppl.* **101,** 9–23.
2. Blinzinger, K. and Kreutzberg, G. (1968) Displacement of synaptic termi- nals from regenerating motoneurons by microglial cells. *Z. Zellforsch* **85,** 145–157.
3. Graeber, M. B., Bise, K., and Mehraein, P. (1993) Synaptic stripping in the human facial nucleus. *Acta Neuropathol.* **86,** 179–181.
4. Stollberg, J. (1995) Synapse elimination, the size principle, and hebbian synapses. *J. Neurobiol.* **26,** 273–282.
5. Chang, Q. and Balice-Gordon, R. J. (1997) Nip and tuck at the neuromus- cular junction: a role for proteases in developmental synapse elimination. *Bioessays* **19,** 271–275.
6. Adams, I. and Jones, D. G. (1982) Synaptic remodelling and astrocytic hyper- trophy in rat cerebral cortex from early to late adulthood. *Neurobiol. Aging* **3,** 179–186.
7. Adams, I. (1987) Comparison of synaptic changes in the precentral and postcentral cerebral cortex of aging humans: a quantitative ultrastructrual study. *Neurobiol. Aging* **8,** 203–212.

8. Terry, R. D., Masliah, E., Salmon, D. P., et al. (1991) Physical basis of cognitive alterations in Alzheimer's disease: synapse loss is the major correlate of cognitive impairment. *Ann. Neurol.* **30,** 572–580.
9. Barron, K. D. (1975) Ultrastructural changes in dendrites of central neurons during axon reaction. *Adv. Neurol.* **12,** 381–398.
10. Obata, H. L., Kubo, S., Kinoshita, H., Murabe, Y., and Ibata, Y. (1981) The effect of small doses of kainic acid on the area CA3 of the hippocampal formation. An electron microscopic study. *Arch. Histol. Jap.* **44,** 135–149.
11. Scheff, S. W. and Price, D. A. (1993) Synapse loss in the temporal lobe in Alzheimer's disease. *Ann. Neurol.* **33,** 190–199.
12. Scheff, S. W., Dekosky, S. T., and Price, D. A. (1990) Quantitative assessment of cortical synaptic density in Alzheimer's disease. *Neurobiol. Aging* **11,** 29–37.
13. Brun, A., Liu, X., and Erikson, C. (1995) Synapse loss and gliosis in the molecular layer of the cerebral cortex in Alzheimer's disease and in frontal lobe degeneration. *Neurodegeneration* **4,** 171–177.
14. Samuel, W. O., Masliah, E., Hill, L. R., Butters, N., and Terry, R. (1994) Hippocampal connectivity and Alzheimer's dementia: effects of synapse loss and tangle frequency in a two-component model. *Neurology* **44,** 2081–2088.
15. Sze, C. I., Troncoso, J. C., Kawas, C., Mouton, P., Price, D. L., and Martin, L. J. (1997) Loss of the presynaptic vesicle protein synaptophysin in hippocampus correlates with cognitive decline in Alzheimer's disease. *J. Neuropathol. Exp. Neurol.* **56,** 933–944.
16. Harigaya, Y., Shoji, M., Shirao, T., and Hirai, S. (1996) Disappearance of actin-binding protein, drebrin, from hippocampal synapses in Alzheimer's disease. *J. Neurosci. Res.* **43,** 87–92.
17. Wakabayashi, K., Honer, W. G., and Masliah, E. (1994) Synapse asterations in the hippocampal-entorhinal formation in alzheimer's disease with and without Lewy body disease. *Brain Res.* **667,** 24–32.
18. Callahan, L. M. and Coleman, P. D. (1995) Neurons bearing neurofibrillary tangles are responsible for selected synaptic deficits. *Neurobiol. Aging* **16,** 311–314.
19. Su, J. H., Deng, G. M., and Cotman, C. W. (1997) Neuronal DNA damage precedes tangle formation and is associated with up-regulation of nitrotyrosine in Alzheimer's disease brain. *Brain Res.* **774,** 193–199.
20. Su, J. H., Satou, T., Anderson, A. J., and Cotman, C. W. (1996) Up-regulation of Bcl-2 is associated with neuronal DNA damage in Alzheimer's disease. *Neuroreport* **7,** 437–440.
21. Masliah, E., Hansen, L., Albright, T., Mallory, M., and Terry, R. D. (1991) Immunoelectron microscopic study of synaptic pathology in Alzheimer's disease. *Acta Neuropathol.* **81,** 428–433.
22. Shimohama, S., Kamiya, S., Taniguchi, T., Akagawa, K., and Kimura, J. (1997) Differential involvement of synaptic vesicle and presynaptic plasma membrane proteins in Alzheimer's disease. *Biochem. Biophys. Res. Commun.* **236,** 239–242.
23. Sze, C. I., Bi, H., Kleinschmidt-DeMasters, B. K., Filley, C. M., and Martin, L. J. (2000) Selective regional loss of exocytotic presynaptic vesicle proteins in Alzheimer's disease brains. *J. Neurol. Sci.* **175,** 81–90.
24. Naito, M., Nagashima, K., Mashima, T., and Tsuro, T. (1997) Phosphatidylserine externalization is a downstream event of interleukin-1β-converting enzyme family protease activation during apoptosis. *Blood* **89,** 2060–2066.

25. Fadok, V. A., Bratton, D. L., Rose, D. M., Pearson, A., Ezekewitz, R. A., and Henson, P. M. (2000) A receptor for phosphatidylserine-specific clearance of apoptotic cells. *Nature* **405,** 85–90.

26. Sambrano, G. R. and Steinberg, D. (1995) Recognition of oxidatively damaged and apoptotic cells by an oxidized low density lipoprotein receptor on mouse peritoneal macrophages: role of membrane phosphatidylserine. *Proc. Natl. Acad. Sci. USA* **92,** 1396–1400.

27. Yang, F., Sun, X., Beech, W., et al. (1998) Antibody to caspase-cleaved actin detects apoptosis in differentiated neuroblastoma and neurons and plaque associated neurons and microglia in Alzheimer's disease. *Am. J. Pathol.* **152,** 379–389.

28. Cole, G. M., Yang, F., Chen, P. P., Frautschy, S. A., Hsiao, K. (1999) Caspase activation in dystrophic neurites in Alzheimer's disease and aged huAPPsw transgenic mice, in *Alzheimer's Disease and Related Disorders* (Iqbal, K., Swaab, D. F., Winblad, B., and Wisniewski, H. M., eds.), Wiley, West Sussex, UK, pp. 363–370.

29. Peters, A., Palay, S. L., and Webster, H. D. (1991) *The Fine Structure of the Nervous System*, Oxford University Press, New York.

30. Mack, A., Furmann, C., and Hacker, G. (2000) Detection of caspase-activation in intact lymphoid cells using standard caspase substrates and inhibitors. *J. Immunol. Meth.* **241,** 19–31.

31. Li, J., Bombeck, C. A., Yang, S., Kim, Y. M., and Billiar, T. R. (1999) Nitric oxide suppresses apoptosis via interrupting caspase activation and mitochondrial dysfunction in cultured hepatocytes. *J. Biol. Chem.* **274,** 17,325–17,333.

32. Schulz, J. B., Weller, M., Matthews, R. T., et al. (1998) Extended therapeutic window for caspase inhibition and synergy with MK-801 in the treatment of cerebral histotoxic hypoxia. *Cell Death Differ.* **5,** 847–857.

33. Teter, B., Yang, F., Wasterlain, C., et al. (1996) Cleavage site-specific cell death switch antibody labels Alzheimer's lesions. *Soc. Neurosci.* **22,** 258.

34. Bozyczko-Coyne, D., Dobrzanski, P., Meyer, S., et al. (1996) ICE-related proteases involved in growth factor deprivation-induced neuronal apoptosis. *Soc. Neurosci.* **22,** 566.

35. Siman, R., Bozycyczko-Coyne, D., Meyer, S., and Bhat, R. V. (1999) Immunolocalization of capsase proteolysis in situ: evidence for widespread caspase-mediated apoptosis of neurons and glia in the postnatal rat brain. *Neuroscience* **92,** 1425–1442.

36. Srinivasan, A., Roth, K. A., Sayers, R. O., et al. (1998) In situ immunodetection of activated caspase-3 in apoptotic neurons in the developing nervous system. *Cell Death Differ.* **5,** 1004–1016.

37. LeBlanc, A., Liu, H., Goodyer, C., Bergeron, C., and Hammond, J. (1999) Caspase-6 role in apoptosis of human neurons, amyloidogenesis, and Alzheimer's disease. *J. Biol. Chem.* **274,** 23,426–23,436.

38. Suurmeijer, A. J. H., van der Wijk, J., van Veldhuisen, D. J., Yang, F., and Cole, G. M. (1999) Fractin immunostaining for the detection of apoptotic cells and apoptotic bodies in formalin-fixed and paraffin-embedded tissue. *Lab. Invest.* **79,** 619–620.

39. Rossiter, J. P., Anderson, L. L., Yang, F., and Cole, G. M. (2000) Caspase-cleaved actin (fractin) immunolabelliing of Hirano bodies. *Neuropathol. Appl. Neurobiol.* **26,** 342–346.

40. Glazner, G. W., Chan, S. L., Lu, C., and Mattson, M. P. (2000) Caspase-mediated degradation of AMPA receptor subunits: a mechanism for preventing excitotoxic necrosis and ensuring apoptosis. *J. Neurosci.* **20,** 3641–3649.

41. Aronica, E., Dickson, D. W., Kress, Y., Morrison, J. H., and Zukin, R. S. (1998) Non-plaque dystrophic dendrites in Alzheimer hippocampus: a new pathological structure revealed by glutamate receptor immunocytochemistry. *Neuroscience* **82,** 979–991.

42. Hsia, A. Y., Masliah, E., McConlogue, L., et al. (1999) Plaque-independent disruption of neural circuits in Alzheimer's disease mouse models. *Proc. Natl. Acad. Sci. USA* **96,** 3228–3233.

43. Irizarry, M. C., McNamara, M., Fedorchak, K., Hsiao, K., and Hyman, B. T. (1997) APP_{sw} transgenic mice develop age-related Aβ deposits and neuropil abnormalities, but no neuronal loss in CA1. *J. Neuropathol. Exp. Neurol.* **56,** 965–973.

44. Steiner, H., Capell, A., Pesold, B., et al. (1998) Expression of Alzheimer's disease-associated presenilin-1 is controlled by proteolytic degradation and complex formation. *J. Biol. Chem.* **273,** 32,322–32,331.

45. Tesco, G., Kim, T. W., Diehlmann, A., Beyreuther, K., and Tanzi, R. E. (1998) Abrogation of the presenilin 1/beta-catenin interaction and preservation of the heterodimeric presenilin 1 complex following caspase activation. *J. Biol. Chem.* **273,** 33,909–33,914.

46. Gervais, F. G., Xu, D., Robertson, G. S., et al. (1999) Involvement of caspases in proteolytic cleavage of Alzheimer's amyloid-β precursor protein and amyloidogenic A beta peptide. *Cell* **97,** 395–406.

47. Lu, D. C., Rabizadeh, S., Chandra, S., et al. (2000) A second cytotoxic proteolytic peptide derived from amyloid β-protein precursor. *Nature Med.* **6,** 397–404.

48. Mattson, M. P., Keller, J. N., and Begley, J. G. (1998) Evidence for synaptic apoptosis. *Exp. Neurol.* **151,** 35–48.

49. Mattson, M. P., Partin, J., and Begley, J. G. (1998) Amyloid β-peptide induces apoptosis-related events in synapses and dendrites. *Brain Res.* **807,** 161–176.

50. Campenot, R. B. (1982) Development of sympathetic neurons in compartmentalized cultures: II. Local control of neurite survival by nerve growth factor. *Devel. Biol.* **93,** 13–21.

51. Campenot, R. B. (1994) NGF and the local control of nerve terminal growth. *J. Neurobiol.* **6,** 599–611.

52. Ivins, K. J., Bui, E. T. N., and Cotman, C. W. (1998) β-Amyloid induces local neurite degeneration in cultured hippocampal neurons: evidence for neuritic apoptosis. *Neurobiol. Dis.* **5,** 365–378.

53. Finn, J. T., Weil, M., Fabienne, A., Siman, R., Srinivasan, A., and Raff, M. C. (2000) Evidence that Wallerian degeneration and localized axon degeneration induced by local neurotrophin deprivation do not involve caspases. *J. Neurosci.* **20,** 1333–1341.

54. Bahr, B. A., Hoffman, K. B., Vanderklish, P. W., et al. (1996) The Alzheimer Aβ1-42 peptide induces CA1 specific Aβ immunostaining and synaptic decay in hippocampal slice cultures. *Soc. Neurosci. 26 Annual* **22,** 480.12.

55. Bahr, B. A., Hoffman, K. B., Yang, A. J., Hess, U. S., Glabe, C. G., and Lynch, G. (1997) Amyloid β-protein is selectively internalized by hippocampal field CA1 and causes neurons to accumulate amyloidogenic carboxyterminal APP fragments. *Neuron* **1,** 1–21.

56. Teter, B., Harris-White, M., Frautschy, S. A., and Cole, G. M. (1999) Role of apolipoprotein E and estrogen in mossy fiber sprouting in hippocampal slice cultures. *Neuroscience* **91**, 1009–1016.
57. Harris-White, M. E., Teter, B., Chu, T., et al. (1998) Effects of ApoE on Aβ deposition in murine organotypic slice cultures. *Soc. Neurosci.* **24**, β849.3.
58. Lambert, M. P., Barlow, A. K., Chromy, B. A., et al. (1998) Diffusible, non-fibrillar ligands derived from Aβ1-42 are potent central nervous system neurotoxins. *Proc. Natl. Acad. Sci. USA* **95**, 6448–6453.
59. Harris-White, M. E., Frautschy, S. A., Cole, G. M. (1998) Methods for evaluating a slice culture model of Alzheimer's disease, in *Methods in Brain Aging* (Timiras, P. and Sternberg, H., eds.), Springer-Verlag, Berlin, pp. 55–65.
60. Butterfield, D. A., Hensley, K., Harris, M., Mattson, M., and Carney, J. (1994) β-Amyloid peptide free radical fragments initiate synaptosomal lipoperoxidation in a sequence-specific fashion: implications to Alzheimer's disease. *Biochem. Biophys. Res. Commun.* **200**, 710–715.
61. Wolf, M. E. and Kapatos, G. (1989) Flow cytometric analysis of rat striatal nerve terminals. *J. Neurosci.* **9**, 94–105.
62. Wolf, M. E. and Kapatos, G. (1989) Flow cytometric analysis and isolation of permeabilized dopamine nerve terminals. *J. Neurosci.* **9**, 106–114.
63. Wolf, M. E., Granneman, J. G., and Kapatos, G. (1991) Characterization of the distribution of G alpha o in rat striatal synaptosomes and its colocalization with tyrosine hydroxylase. *Synapse* **9**, 66–74.
64. Wolf, M. E., LeWitt, P. A., Bannon, M. J., Dragovic, L. J., and Kapatos, G. (1991) Effect of aging on tyrosine hyrdroxylase protein content and the relative number of dopamine nerve terminals in human caudate. *J. Neurochem.* **56**, 1191–1200.
65. Gylys, K. H., Fein, J. A., and Cole, G. M. (2000) Quantitative characterization of crude synaptosomal fraction (P-2) components by flow cytometry. *J. Neurosci. Res.* **61**, 186–192.
66. Dodd, P. R., Hardy, J. A., Baig, E. B., et al. (1986) Optimization of freezing, storage, and thawing conditions for the preparation of metabolically active synaptosomes from frozen rat and human brain. *Neurochem. Pathol.* **4**, 177–198.
67. Gylys, K. H., Fein, J. A., and Cole, G. M. (2000) In vitro activation of caspases in nerve terminals. *Age* **23**, 65.
68. Wolf, B. B., Goldstein, J. C., Stennicke, H. R., et al. (1999) Calpain functions in a caspase-independent manner to promote apoptosis-like events during platelet activation. *Blood* **94**, 1683–1692.
69. Mark, R. J., Hensley, K., Butterfield, D. A., and Mattson, M. P. (1995) Amyloid β-peptide impairs ion-motive ATPase activities: evidence for a role in loss of neuronal Ca^{2+} homeostasis and cell death. *J. Neurosci.* **15**, 6239–6249.
70. Lankiewicz, S., Luetjens, C. M., Bui, N. T. K. A. J., et al. (2000) Activation of calpain I converts excitotoxic neuron death into a caspase-independent cell death. *J. Biol. Chem.* **275**, 17,064–17,071.
71. Squier, M. K., Sehnert, A. J., Sellins, K. S., Malkinson, A. M., Takano, E., and Cohen, J. J. (1999) Calpain and calpastatin regulate neutrophil apoptosis. *J. Cell. Physiol.* **178**, 311–319.
72. Squier, M. K. and Cohen, J. J. (1999) Calpain, an upstream regulator of thymocyte apoptosis. *J. Immunol.* **178**, 311–319.
73. Wang, K. K. W. (2000) Calpain and caspase: can you tell the difference? *Trends Neurosci.* **23**, 20–26.

74. Cheng, A. G., Huang, T., Stracher, A., et al. (1999) Calpain inhibitors protect auditory sensory cells from hypoxia and neurotrophin-withdrawal induced apoptosis. *Brain Res.* **850,** 234–243.

75. Rami, A., Agarwal, R., Botez, G., and Winckler, J. (2000) mu-calpain activation, DNA fragmentation, and synergistic effects of caspase and calpain inhibitors in protecting hippocampal neurons from ischemic damage. *Brain Res.* **866,** 299–312.

76. Masliah, E., Iimoto, D., Saitoh, T., Hansen, L. A., and Terry, R. D. (1990) Increased immunoreactivity of brain spectrin in Alzheimer disease: a marker for synapse loss? *Brain Res.* **531,** 36–44.

77. Lee, M. S., Kwon, Y. T., Li, M., Peng, J., Friedlander, R. M., and Tsai, L. H. (2000) Neurotoxicity induces cleavage of p35 to p25 by calpain. *Nature* **405,** 360–364.

78. Grynspan, F., Griffin, W. R., Cataldo, A., Katayama, S., and Nixon, R. A. (1997) Active site-directed antibodies identify calpain II as an early-appearing and pervasive component of neurofibrillary pathology in Alzheimer's disease. *Brain Res.* **763,** 145–158.

79. Beckman, K. B. and Ames, B. N. (1998) The free radical theory of aging matures. *Physiol. Rev.* **78,** 548–581.

80. Perry, G. and Smith, M. A. (1997) A central role for oxidative damage in the pathogenesis and therapeutics of Alzheimer's disease. *Alzheimer's Res.* **2,** 319–324.

81. Brenner, C., Marzo, I., and Kroemer, G. (1998) A revolution in apoptosis: from a nucleocentric to a mitochondriocentric perspective. *Exp. Gerontol.* **33,** 543–553.

82. Keller, J. N., Pang, Z., Geddes, J. W., et al. (1997) Impairment of glucose and glutamate transport and induction of mitochondrial oxidative stress and dysfunction in synaptosomes by amyloid β-peptide: role of the lipid peroxidation product 4-hydroxynonenal. *J. Neurochem.* **69,** 273–284.

83. Behl, C., Davis, J. B., Lesley, R., and Schubert, D. (1994) Hydrogen peroxide mediates amyloid β protein toxicity. *Cell* **77,** 817–827.

84. Butterfield, D. A., Martin, L., Carney, J. M., and Hensley, K. (1996) Aβ(25-35) peptide displays H202-like reactivit towards aqueous Fe^{2+}, nitroxide spin probes, and synaptosomal membrane proteins. *Life Sci.* **58,** 217–228.

85. Rogers, J., Cooper, N. R., Webster, S., et al. (1992) Complement activation by β-amyloid in Alzheimer disease. *Proc. Natl. Acad. Sci. USA* **89,** 10,016–10,020.

86. Webster, S., Lue, L. F., Brachova, L., et al. (1997) Molecular and cellular characterization of the membrane attack complex, C5b-9, in Alzheimer's disease. *Neurobiol. Aging* **18,** 415–421.

87. Yang, A. J., Knauer, M., Burdick, D. A., and Glabe, C. (1995) Intracellular Aβ1-42 aggregates stimulate the accumulation of stable, insoluble amyloidogenic fragments of the amyloid precursor protein in transfected cells. *J. Biol. Chem.* **270,** 14,786–14,792.

88. Yang, A. J., Chandswangbhuvana, D., Margol, L., and Glabe, C. G. (1998) Loss of endosomal/lysosomal membrane impermeability is an early event in amyloid Aβ1-42 pathogenesis. *J. Neurosci. Res.* **52,** 691–698.

89. Deiss, L. P., Galinka, H., Berissi, H., Cohen, O., and Kimchi, A. (1996) Cathepsin D protease mediates programmed cell death induced by interferon-γ, Fas?APO-1 and TNF-a. *EMBO J.* **15,** 3861–3870.

In Situ Detection
of Apoptotic Neurons

Kevin A. Roth

1. Introduction

Neuronal apoptosis plays a significant role in nervous system development and in neuropathological conditions *(1)*. Although apoptosis is not the only type of non-necrotic, regulated cell death observed in the nervous system, it is by far the most extensively investigated *(2,3)*. Studies in a variety of species, cell types, and experimental paradigms have led to a better understanding of the molecular pathways that regulate apoptosis and recognition of the complexities of the apoptotic process. Apoptotic cell death was originally defined based on cytological features including cell shrinkage, chromatin condensation and margination, nuclear pyknosis and fragmentation, cytoplasmic membrane blebbing, and formation of apoptotic bodies *(4)*. These features, which are best appreciated by ultrastructural examination, remain the gold standard by which any given cell can be classified as "apoptotic." Unfortunately, it is impractical to perform detailed electron microscopic analyses on a large number of samples on a routine basis. For this reason, a variety of histological and immunohistochemical techniques have been developed to detect apoptotic cells *in situ*. These methods are based on light microscopic visualization of apoptotic cytological features, immunohistochemical detection of apoptosis-associated molecules, and/or demonstration of apoptosis-related cellular events *(5)*.

2. Molecular Characterization of Neuronal Apoptosis

The molecular pathway regulating apoptosis has been largely conserved throughout evolution. In the nematode *Caenorhabditis elegans*, apoptosis is controlled by a genetic pathway involving

From: *Neuromethods, Vol. 37: Apoptosis Techniques and Protocols, 2nd Ed.*
Edited by: A. C. LeBlanc @ Humana Press, Inc., Totowa, NJ

the coordinated action of EGL-1, CED-9, CED-4, and CED-3 (6). Structural homologs of these molecules exist in mammals and consist of BH3-only Bcl-2 subfamily members, anti-apoptotic Bcl-2 subfamily members, APAF-1-like molecules, and the caspase family, respectively (7,8). Bcl-2 family members affect a variety of intracellular processes, most important of which is mitochondrial function, including cytochrome-c subcellular localization. Apoptotic stimuli, typically through a Bcl-2 family-mediated checkpoint, cause redistribution of cytochrome-c from mitochondria to cytosol (7). Cytosolic cytochrome-c binds to APAF-1 in the presence of dATP or ATP, resulting in the recruitment and activation of caspase-9 and subsequent cleavage of caspase-3. Caspase-3 is considered an effector caspase and its activation results in the cleavage of a variety of intracellular targets and, ultimately, extensive DNA fragmentation. In most cell types including neurons, caspase-3 deficiency and/or pharmacological inhibition of caspase activity markedly attenuates or prevents occurrence of the apoptotic phenotype (9). Caspase inhibition may or may not, however, prevent cell death, depending on the apoptotic stimulus, the cell type, and the death commitment point.

My laboratory has utilized mice with targeted gene disruptions in specific apoptosis-associated molecules to investigate the molecular regulation of neuronal cell death (1). These studies, and those of other laboratories, have demonstrated important roles for caspase-9, APAF-1, and caspase-3 in regulating neural precursor cell death and Bcl-X_L and Bax, along with caspase-9, APAF-1, and caspase-3, in regulating neuron survival during embryonic development. These and other apoptosis-associated molecules including Bcl-2, Bak, caspase-2, caspase-6, and caspase-8 also affect neuronal apoptosis in the mature nervous system and in a variety of experimental paradigms, indicating that Bcl-2 and caspase family members continue to play important roles in regulating neuron survival throughout adult life. Definitive evidence of a role for specific apoptosis-associated molecules in human neuropathological conditions is more difficult to obtain, but neuronal apoptosis has been implicated in a variety of neurological diseases (10). During the course of our investigations of neuronal apoptosis, we have used and/or developed a variety of techniques to identify apoptotic cells in the developing and mature nervous system. These techniques, particularly when used in combination, can provide strong support for classifying neuronal death as "apoptotic" and may give insights into the underlying molecular regulation of neuronal cell death.

3. Histochemical Detection of Apoptotic Neurons

The first step in any investigation of neuronal apoptosis necessarily involves careful histological examination of tissue sections to identify apoptotic neurons *(11,12)*. By definition, apoptotic cells must have the characteristic cytological features originally used to define the term "apoptosis." Also implicit in this definition is that apoptotic neurons are nonviable and their corpses will be rapidly eliminated through phagocytosis. With rare exceptions, neurogenesis is largely restricted to the developing nervous system and, thus, the consequence of neuronal apoptosis in the adult nervous system is neuron loss. Apoptosis is relatively easy to detect in the developing nervous system, where programmed cell death predictably occurs in specific neuroanatomical sites and cell subpopulations. Similarly, acute neurotoxic injury in the adult brain may affect large subpopulations of neurons in a relatively synchronous fashion. More difficult to detect, however, is neuronal apoptosis in the setting of chronic injury or neurodegenerative conditions, when neurons might be expected to undergo apoptosis in a sporadic or asynchronous fashion. No single histological method is applicable to all investigations, and therefore I will describe several techniques that we have used to identify apoptotic neurons in brain sections from both humans and experimental animals.

Short of electron microscopic examination, the most reliable method for detection of apoptotic cytological features in neurons is light microscopic examination of optimally fixed and stained plastic-embedded tissue sections with a high-power objective. In experimental adult animals, this is accomplished by perfusing anesthetized rats or mice with heparinized saline followed by 3% glutaraldehyde in 0.1 M phosphate buffer (pH 7.3) containing 0.45 mmol/L Ca^{2+} or with other glutaraldehyde-based fixatives. The brain is then removed from the skull and fixed for an additional 24–48 h before further dissection, dehydration in graded ethanol solutions, and embedding in Epon with propylene oxide as an intermediary solvent. One-micrometer-thick plastic sections are cut and then stained with toluidine blue. For embryonic brains or appropriately small (1–2 mm thick) samples of human brain, immersion fixation in glutaraldehyde fixative for 48–72 h is typically adequate to achieve well-preserved specimens.

Figure 1A shows a toluidine blue-stained 1-μm-thick section from the brainstem of an embryonic d 12 mouse that lacks Bcl-X$_L$. We have shown that targeted disruption of *bcl-x* results in a dramatic increase in apoptosis throughout the immature neuronal population in the developing nervous system *(13)*. Cells at various stages of apoptotic degeneration can be seen interspersed with viable neurons. Apoptotic neurons can also be readily identified in sections from wild-type embryos at known sites of programmed cell death (data not shown). Possible drawbacks to this method of apoptotic neuron detection are that glutaraldehyde fixation and plastic embedding may destroy antigen reactivity, precluding additional immunohistochemical analysis of these tissues and the technically challenging processing and preparation of 1-μm thick sections in inexperienced laboratories.

Adequate detection of cytologically apoptotic neurons can usually be obtained from fixed paraffin-embedded brain sections cut at a thickness of 3–5 μm. In our laboratory, we prefer Bouin's fixative (five parts formaldehyde, 15 parts saturated picric acid; one part glacial acetic acid) to the more commonly used 4% paraformaldehyde fixation, since it provides better nuclear detail in histochemically stained sections. For small samples, e.g., mouse embryos, immersion fixation for 12–24 h in Bouin's solution at 4°C is optimal. Because of the acidity of the fixative, perfusion with Bouin's is difficult; however, it can be used as a postfixative following 4% paraformaldehyde perfusion of rodent brains. Alternatively, brains can be cut into multiple blocks and immersion-fixed as with whole embryos. Routine hematoxylin and eosin staining of sections is usually adequate to demonstrate apoptotic cells, although other histochemical stains may also be used to detect apoptosis *(14)*. Figure 1B shows apoptotic neurons in the Bcl-X$_L$-deficient embryonic brain identified by hematoxylin and eosin staining. Finally, fluorescent dyes such as DAPI, Hoechst 33,258, and propidium iodide bind to nucleic acids and are sensitive indicators of apoptotic nuclear morphology. Figure 2A demonstrates condensed, marginated, and/or fragmented nuclear features in apoptotic cortical neurons in a Hoechst 33,258-stained section (1 μg/mL for 5 min) from a neonatal mouse exposed to ethanol. Such fluorescent dyes are particularly useful when used in combination with immunohistochemical detection and/or TUNEL techniques (see Subheading 7.4.).

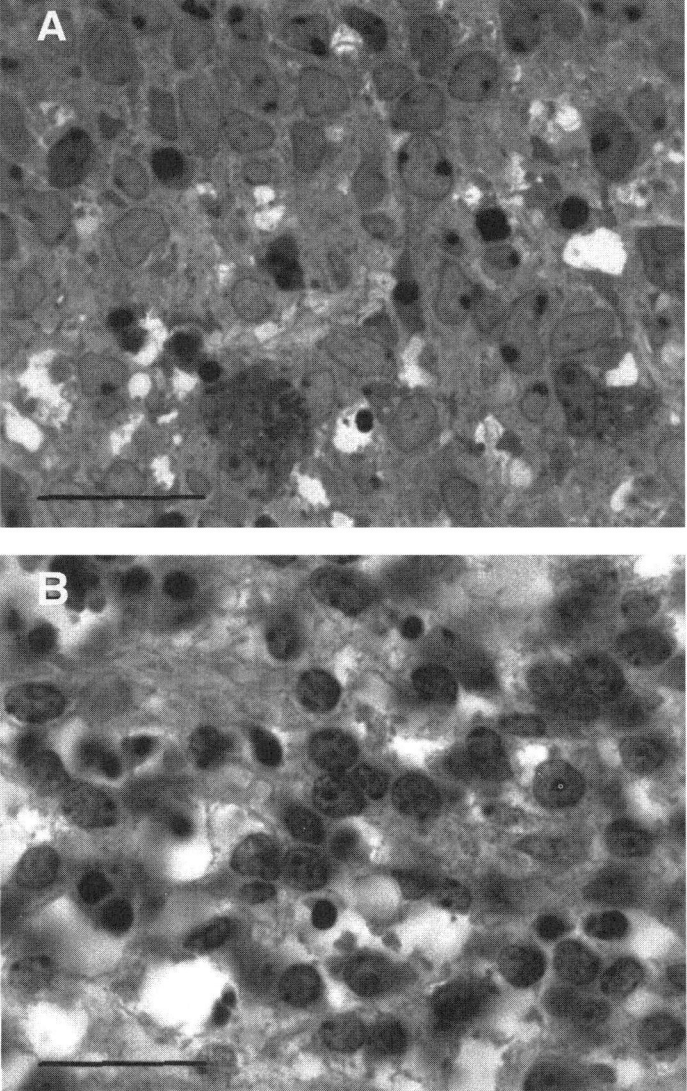

Fig. 1. Light microscopic detection of apoptotic neurons. Apoptotic neurons in the brainstem of embryonic d 12 Bcl-X$_L$-deficient mice were identified in both toluidine blue-stained, glutaraldehyde-fixed, plastic-embedded, 1 μm-thick sections (**A**) and hematoxylin and eosin-stained Bouin's solution-fixed, paraffin-embedded, 4-μm-thick sections (**B**). Numerous cells with condensed, marginated, and/or fragmented chromatin consistent with apoptotic nuclear changes are identified. Scale bars equal 20 μm.

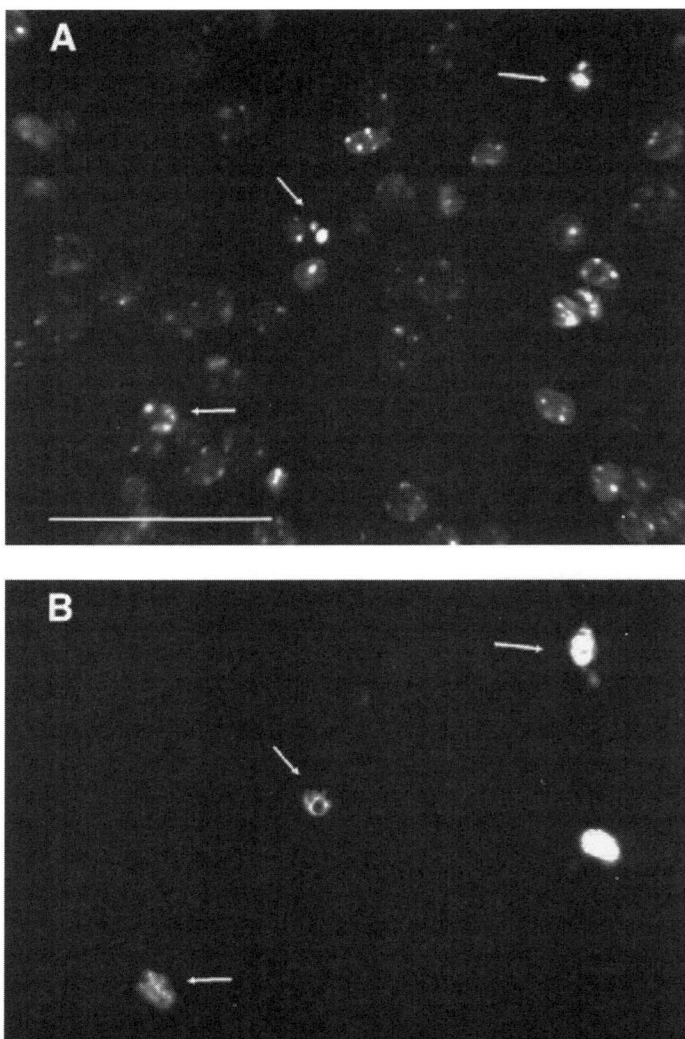

Fig. 2. Fluorescence detection of apoptotic nuclei and activated caspase-3 immunoreactivity. A section from the brain of a postnatal d 7 mouse exposed to ethanol 8 h previously was stained with Hoechst 33,258 (**A**) and immunolabeled with antibodies against activated caspase-3 (**B**). Many cortical neurons show condensed and fragmented nuclear features simultaneously with activated caspase-3 immunoreactivity, consistent with apoptosis (examples are indicated by arrows). Scale bars equal 50 μm.

4. Immunohistochemical Detection of Apoptosis-Associated Molecules

Molecular characterization of apoptotic death pathways has provided new targets for which to aim in our quest for sensitive *in situ* apoptosis detection. The apoptotic nuclear phenotype typically requires caspase activity for its occurrence *(15,16)*. Caspases 3, 6, and 7 are the best-defined effector caspases, and targeted disruption of caspase-3 has been shown to have a marked effect on nervous system development and neuronal apoptosis *(17)*. Caspases are cysteine proteases that exist in cells at baseline as inactive zymogens *(8)*. Upon receipt of an apoptotic stimulus, the caspase zymogen becomes cleaved at specific aspartic residues and generates large and small subunits that together constitute the active caspase. Because of this cleavage event, antibodies can be generated against the carboxyl terminus of the large caspase subunit that will not recognize the inactive zymogen form of the caspase. A number of antibodies have now been generated against various active caspases, including caspases 3, 8, and 9 *(18–20)*. Immunohistochemical studies of neuronal apoptosis have been performed with several of these antibodies, and they have proven to be sensitive indicators of caspase activation and neuronal apoptosis. We have used antibodies against the carboxyl terminus of the large subunit of caspase-3 to detect caspase-3 activation in both neuropathological conditions and normal nervous system development (Fig. 2B). More recently, identification of downstream caspase substrates has led to the development of antibodies against specific caspase-cleaved proteins including fodrin, actin, and poly (ADP-ribose) polymerase *(21–23)*. Some of these antibodies have been used to infer caspase activation in the nervous system. An important caveat to the detection of apoptosis-associated caspase cleavage is that, in and of itself, it is insufficient to classify a cell or process as apoptotic. To do so requires additional evidence of apoptotic cytological features and cell death.

Cleaved caspases and their downstream substrates represent attractive targets for *in situ* detection of apoptosis-associated events since they are not typically present in neurons at baseline. Thus their immunodetection in neurons suggests, but does not prove, that a positive cell is in the process of undergoing apoptosis. More difficult to demonstrate in tissue sections but equally associated with apoptosis is a change in the subcellular localization of Bcl-2 family members and cytochrome-*c*. Translocation of Bax from cytosol to mitochondria is an important trigger for release of cytochrome-*c* from mitochondria to cytosol. Translocation of Bax and/or cytochrome-*c* in various experimental apoptotic paradigms has been immunohistochemically demonstrated *(24)*. Such studies require multilabel immunohistochemical detection so that the intracellular localization of Bax and/or cytochrome-*c* can be determined in relationship to well-defined subcellular markers; laser scanning confocal microscopy may be required to achieve adequate signal resolution.

Regardless of which apoptosis-associated molecule one wishes to localize, the keys to successful immunohistochemical detection are appropriate tissue fixation, specific antibodies, and sensitive signal amplification. Unfortunately, there is no single fixative or fixation protocol that will work for all antigens and antibodies. If an immunolabeling protocol for the antibody of interest has been published or recommended by the manufacturer, it is usually best to follow the established protocol. When testing new antibodies, we typically begin with paraffin-embedded sections and compare three different tissue fixatives, 4% paraformaldehyde (in 0.01 *M* phosphate-buffered saline, PBS, pH = 7.2), Bouin's solution, and methacarn (6 parts methanol: 3 parts chloroform: 1 part glacial acetic acid). Paraformaldehyde and Bouin's fixed tissue will also be examined without and with heat-induced epitope retrieval (0.01 *M* citrate buffer, pH 6.0, for 20 min in a rice steamer or pressure cooker, followed by a 20-min cooldown). If adequate results are not obtained, we will test cryostat sections prepared from 4% paraformaldehyde-fixed and cryopreserved (30% sucrose in PBS for 24–48 h at 4°C) brain. Alternatively, fresh-frozen brain sections are prepared and fixed in one of several precipitating fixatives such as ethanol, methanol, or acetone (air dried cryosections are placed in the fixative at −20°C for 10–20 min, followed by air drying). If fixation conditions cannot be altered, e.g., in a retrospective study of human brain sections obtained at surgical resection or autopsy, antigen retrieval techniques may be necessary to achieve adequate immunostaining results. Numerous antigen retrieval solutions are commercially available and may be critical for immunodetection in overly fixed tissue.

There are many different methods for localizing antibody binding in tissue sections. The investigator can decide between chromogenic or fluorescence detection depending on personal preference, equipment availability, and experimental design. The simplest method for localizing antibody binding is to use secondary antibodies that are either fluorescently labeled or conjugated to an enzyme such as horseradish peroxidase (HRP), which can then be used to deposit a chromogenic substrate, e.g., diaminobenzidine (DAB). Although simple, these methods lack sensitivity. To dramatically increase immunohistochemical detection sensitivity, we have made extensive use of tyramide signal amplification procedures (TSA; Perkin-Elmer Life Science Products, Boston, MA). In TSA, HRP is used to deposit labeled tyramine at sites of antibody binding *(25,26)*. The label can consist either of a fluorophore (TSA direct) or an intermediate molecule such as biotin, which is subsequently detected with additional reagents (TSA indirect). Unlike traditional precipitating HRP substrates, tyramine is converted by HRP into a highly reactive oxidized intermediate that binds rapidly and covalently to cell-associated proteins at sites of HRP activity. TSA produces a 10- to 100-fold increase in signal intensity compared to standard immunohistochemical detection methods *(27)*. A second generation of TSA reagents called "TSA Plus" (Perkin-Elmer Life Science Products) provides even greater signal amplification and has proven extremely useful for mRNA *in situ* hybridization *(28)*. Multilabel TSA detection can be easily performed and permits a sensitive analysis of the relationship between apoptosis-associated molecules and/or apoptosis-associated cytological changes. Our detailed protocol for TSA-based immunohistochemistry is provided under Subheading 7.

5. Detection of Apoptosis-Associated Cellular Events

The apoptotic process produces a variety of physical and biochemical changes within a cell, including the redistribution of phosphatidylserine from the inner to the outer leaflet of the plasma membrane lipid bilayer and extensive DNA fragmentation *(29,30)*. Phosphatidylserine externalization can be detected by annexin-V binding, and this technique has proven to be a sensitive indicator of apoptosis in vitro. Unfortunately, in tissue sections, annexin-V labeling does not specifically label externalized phosphatidylserine, since tissue processing and sectioning permits annexin-V access to phosphatidylserine on the inner membrane surface. In-vivo perfusion of animals with labeled annexin-V followed by in-vitro detection of bound annexin-V has been used successfully to detect apoptotic cells in tissue sections *(31)*; however, this technique has limited applications.

Biochemical studies have demonstrated that apoptotic cells typically show increased endonuclease activity and cleavage of their DNA into oligonucleosomal fragments that produce a "ladder" pattern on agarose gel electrophoresis *(32)*. Methods for detection of these DNA strand breaks and other apoptosis-associated biochemical changes in chromatin structure have been developed and are widely popular for *in situ* detection of apoptotic neurons *(33)*. Several investigators have developed antibodies against single-stranded DNA and we, and others, have used them to detect apoptotic cells in the developing nervous system and in a variety of experimental paradigms *(34,35)*. These antibodies can bind to relatively short stretches of single-stranded DNA, including those generated during formation of oligonucleosomal ladders. For example, Fig. 3A illustrates numerous single-stranded DNA antibody immunoreactive cells in the embryonic ventricular zone following intrauterine exposure to cytosine arabinoside. These same cells show apoptotic nuclear features and activated caspase-3 immunoreactivity (data not shown). Recently, Frankfurt and Krishan described a method for *in situ* detection of apoptotic cells that utilized a monoclonal antibody against single-stranded DNA and formamide-induced DNA denaturation of condensed chromatin *(36)*. These antibodies require a long stretch of single-stranded DNA for binding and do not detect single-stranded DNA ends that may contribute to nonspecific labeling using other single-stranded DNA antibodies and/or terminal deoxynucleotidyl transferase (TdT)-mediated dUTP nick end labeling (TUNEL) detection. Using a combination of heat and formamide, apoptotic nuclei were specifically recognized in both frozen and paraffin-embedded sections by their monoclonal antibody, presumably as a result of altered DNA–histone interactions in the apoptotic cells. Additional studies are required to determine the specificity and sensitivity of this method for *in situ* detection of neuronal apoptosis.

Apoptotic nuclear changes can also be detected *in situ* with TdT-dependent incorporation of labeled dUTP onto the 3'OH ends of fragmented DNA that are generated during the apoptotic process (Fig. 3B). The major caveat to TUNEL detection is that it may lack specificity, i.e., it may label necrotic cells and/or cells that are not obviously undergoing apoptosis *(37–39)*. TUNEL detection is significantly affected by fixation conditions, tissue type,

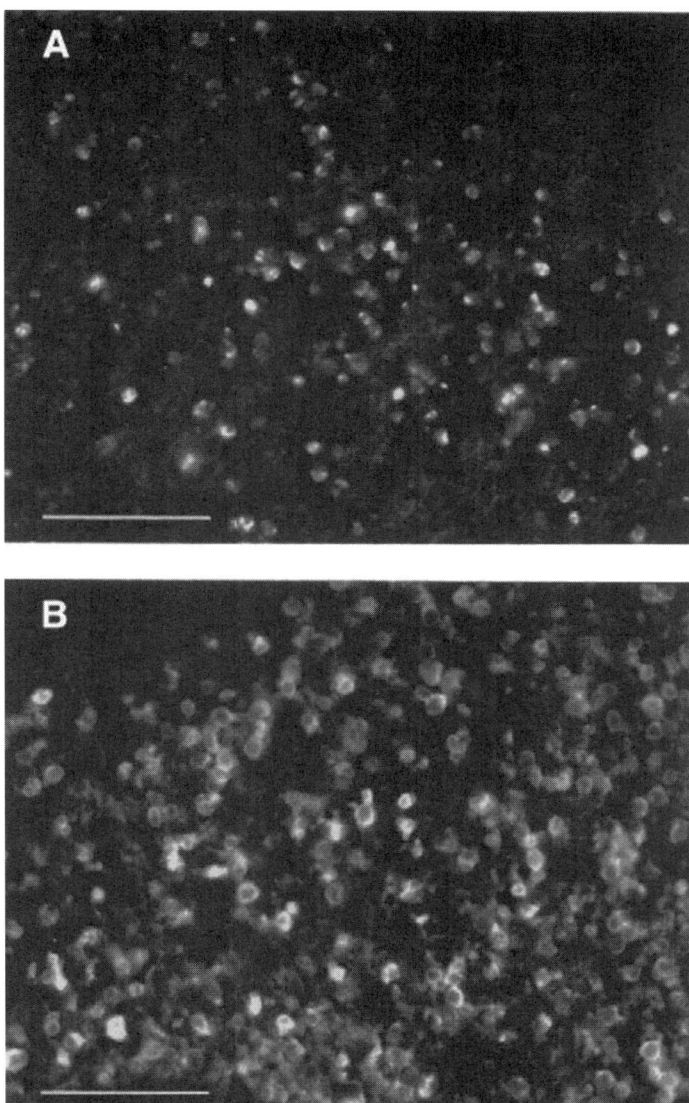

Fig. 3. Detection of apoptosis-associated DNA fragmentation. Sections of brains from embryonic d 12 mice, 8 h posttransplacental exposure to cytosine arabinoside, were immunostained with antibodies against single-stranded DNA (**A**) or labeled using a modified TUNEL protocol (**B**). Extensive labeling of apoptotic neural precursor cells is observed in the ventricular zone. Scale bars equal 50 µm.

reaction conditions, and method of dUTP detection *(40)*. Despite these caveats, sensitive protocols for *in situ* detection of fragmented DNA are available, including protocols for the simultaneous detection of TUNEL-positive nuclei and immunoreactivity for apoptosis-associated molecules (*[41]*, see Subheading 7.4.). The multilabeling technique that we developed permits simultaneous detection of TUNEL positivity, apoptotic nuclear features (Hoechst 33,258 staining), and activated caspase immunoreactivity. Under such labeling conditions, a cell can confidently be classified "apoptotic." Finally, a variety of other techniques have been described to detect apoptotic cells *in situ*, and these may have significant utility in some experimental paradigms *(42–44)*.

6. Conclusions

Neuronal apoptosis is the focus of intense scientific investigation. The ability to detect and accurately classify apoptotic neurons in tissue sections is essential for defining the molecular pathways of neuronal cell death and for determining the significance of this process in nervous system development and neuropathological conditions. By using a combination of techniques to identify apoptosis-associated molecules, cellular events, and cytological changes, apoptotic neurons can be readily identified *in situ*. This approach, particularly when complemented by careful ultrastructural studies, represents an effective and sensitive means for analysis of neuronal apoptosis.

7. Detailed Methods

7.1. Tyramide Signal Amplification (TSA) Immunohistochemical Protocol

1. Prepare sections.
   a. Frozen sections: remove sections from freezer, set at room temperature for 3–5 min, rehydrate in PBS (10 mM, pH 7.2; 8.0 g NaCl, 0.2 g KCl, 1.44 g $Na_2HPO_4 \cdot 2H_2O$ and 0.2 g HK_2PO_4 in 1 L of dH_2O).
   b. Paraffin-embedded sections: deparaffinize (2 × 10 min in xylene, or xylene substitute, e.g., Hemo-De; 3 × 3 min in isopropanol, or graded ethanol dilutions; 3 × 3 min in water).

2. If necessary, perform antigen retrieval. We use various antigen retrieval methods. One method is to incubate sections in 0.01 *M* citrate buffer, pH 6.0, for 20 min in a rice steamer (or in gently boiling water), followed by a 20-min cooldown.

3. If necessary, reduce endogenous peroxidase activity by incubating sections in a H_2O_2 solution. The concentration of H_2O_2 and the length of incubation will depend on the amount of endogenous peroxidase activity present in your tissue and tissue integrity. I typically use room-temperature incubation in either 0.3% H_2O_2 in PBS for 30 min or 3.0% H_2O_2 in PBS for 5 min. The H_2O_2 can be diluted in water or methanol instead of PBS if desired. Following the H_2O_2 incubation, wash sections 3 × 3 min in water (or place in running tap water for 5 min) and for 3 min in PBS.

4. Incubate sections in PBS-BB (PBS with 1% bovine serum albumin, 0.2% powdered skim milk, and 0.3% Triton X-100) for 30 min to block nonspecific antibody binding.

5. Dilute primary antibody in PBS-BB and apply to sections. Incubate sections at 4°C overnight or for 1 h at room temperature.

6. Wash 3 × 5 min in PBS.

7. Apply horseradish peroxidase (HRP)-conjugated secondary antibody diluted in PBS-BB for 1 h at room temperature. Alternatively, apply biotin-conjugated secondary antibody for 1 h, wash 3 × 5 min in PBS, and apply HRP-conjugated streptavidin in PBS-BB for 30 min. (Similarly, ABC methods may be substituted here as well.)

8. Wash 3 × 5 min in PBS.

9. TSA

   a. Direct TSA: Dilute fluorescent tyramide conjugate to the manufacturer's recommended dilution in amplification diluent. Incubate on slides for 3–10 min. Prolong the TSA deposition step for thicker sections. We use 30- to 60-min deposition times for free-floating brain sections (100 μm).

   b. Indirect TSA: Deposit biotin-conjugated tyramide as described in step A. Wash 3 × 5 min in PBS. For fluorescence detection of the deposited biotinyl tyramide, incubate sections for 30 min in fluorophore-conjugated streptavidin diluted in PBS-BB. For chromogenic detection, incubate sections for 30 min in enzyme (HRP or alkaline phosphatase)-conjugated streptavidin, wash 3 × 5 min in PBS, and perform appropriate chromogen deposition.

10. Wash 3×5 min in PBS and cover slip with appropriate mounting medium.

7.1.1. Notes

1. If you use biotin-conjugated secondaries and/or indirect TSA with biotinyl-tyramide deposition, it may be necessary to block endogenous biotin in your tissue. If so, endogenous biotin masking is performed between steps 3 and 4.
2. Tris buffer may be substituted for PBS.
3. Tween-20 (0.05%) may be substituted for Triton X-100 (0.3%) in the blocking buffer.
4. No detergent should be added to the blocking buffer if you are trying to detect membrane-bound antigens in frozen sections.

7.2. Sequential Fluorescent TSA Detection Using Two Unlabeled Primary Antibodies Raised in Separate Species

Co-incubate sections with both primary antibodies and detect the first antigen by following steps 1–10 of the standard TSA protocol.

1. Destroy the activity of the HRP-conjugated secondary antibody used to perform the initial TSA deposition. This can be done by boiling the sections for 1 min in H_2O or incubating them for 30 min in 0.3% H_2O_2 or 5 min in 3.0% H_2O_2. Alternatively, dilute HCl or NaOH can be used to destroy residual HRP activity.
2. Wash 3×5 min in PBS.
3. Repeat steps 5–10 of the standard TSA protocol for the second primary antibody.

7.2.1. Notes

1. Obviously, the two TSA detections must be performed with different fluorophores, i.e., fluorescein and rhodamine, coumarin and rhodamine, fluorescein and cyanine-3, etc., to distinguish the two immunoreactivities.
2. A control section in which the second primary is omitted should be included to ensure adequate destruction of the HRP activity in step 1.
3. The simplest way to perform this protocol is to use HRP-conjugated secondaries and direct TSA detection for both primary antibodies. This minimizes potential crossreactivity between the two TSA detections.

4. If required to increase signal intensity, biotinylated secondaries and/or indirect TSA detection can be performed for one or both of the primary antibodies. However, this may necessitate the use of excess avidin and biotin to mask unreacted biotin molecules between detection steps and requires additional control sections, thus complicating the protocol.

7.3. Detection of Two Unlabeled Primary Antibodies Raised in a Single Species by Sequential TSA and Conventional Fluorescence Detection

For a detailed description of this method, *see* Shindler and Roth, *J. Histochem. Cytochem.* 1996; **44**:1331–1335. Briefly, this protocol is predicated on the fact that TSA dramatically decreases the antibody concentration required to detect antigen. Thus, a primary antibody dilution can be found that is below the detection limit of conventional fluorescence detection but is sufficient for TSA detection. Performance of TSA detection with this concentration of an initial primary antibody permits the subsequent application of a second primary antibody raised in the same species as the first without any detectable recognition of the first primary antibody by the second conventional fluorescence detection system; i.e., because of the low concentration of the first primary antibody it is effectively neglected by the conventional fluorescence detection system used to detect the second primary antibody. To perform this method, you must first experimentally determine a concentration of the first primary antibody that will generate a signal with TSA but not conventional fluorescence techniques. Typically, primary antibody concentrations can be decreased 10 to 100 times for TSA compared to detection with fluoresceinated secondary antibodies.

Using the first primary antibody at an appropriate concentration, detect the first antigen by following steps 1–10 of the standard TSA protocol.

1. Dilute the second primary antibody in PBS-BB and apply to sections. Incubate sections at 4°C overnight or for 1 h at room temperature.
2. Wash 3 × 5 min in PBS.
3. Apply fluorescently conjugated secondary antibody diluted in PBS-BB for 1 h at room temperature.

4. Wash 3 × 5 min in PBS and cover slip with PBS/glycerol (1/1) or other appropriate mounting medium.

7.3.1. Notes

1. Obviously, the fluorophores used to detect the two primary antibodies must be distinct.
2. A control section in which the second primary is omitted should be included to ensure that the first primary is sufficiently diluted to escape detection during step 3.

7.4. Tyramide Signal Amplification of TUNEL Staining

The TSA/TUNEL protocol is described by Shindler, Latham, and Roth, *J. Neurosci.* 1997; **17**:3112–3119, and was modified from a TUNEL protocol described by Tornusciolo, Schmidt, and Roth, *BioTechniques* 1995; **19**:800–805. The protocol described below is designed for paraffin-embedded sections but can be readily modified for use with frozen sections and cells in tissue culture.

7.4.1. TUNEL Buffers

1. TDT buffer (pH 7.2; 100 mL): 30 mM Tris-base (0.36 g), 140 mM cacodylic acid (3.0 g), 1 mM cobalt chloride (23 mg).
2. Stop buffer (100 mL): 300 mM NaCl (1.76 g), 30 mM Na-citrate (0.88 g).
3. TDT reaction mix: digoxygenin-conjugated dUTP (0.25 nmol/ 100 mL TDT buffer), terminal deoxynucleotidyl transferase (25 U/100 mL TDT buffer), TDT buffer.

7.4.2. TSA TUNEL Protocol

1. Deparaffinize appropriately fixed tissue sections.
2. Incubate sections in 0.5% Triton X-100 (in PBS) for 10 min at room temperature.
3. Wash 3 × 3 min in PBS.
4. Incubate sections in TDT buffer for 3 min.
5. Tip off TDT buffer, apply TDT reaction mix, and incubate for 60–90 min at 37°C. The reaction time can be altered to maximize the signal-to-noise ratio.)
6. Wash 3 × 5 min in stop buffer at room temperature.
7. Wash 3 × 3 min in PBS.
8. Incubate sections with HRP-conjugated anti-digoxygenin antibodies (diluted in PBS-BB) overnight at 4°C, or for 1 h at room temperature.

9. Wash 3 × 5 min in PBS.
10. Direct TSA: *see* step 9 of standard TSA protocol.
11. Wash 3 × 5 min in PBS and cover slip with appropriate mounting medium.

7.4.3. Notes

1. Fluorescein conjugated dUTP and anti-fluorescein HRP or biotin-conjugated dUTP and streptavidin HRP can be used instead of digoxygenin dUTP and anti-digoxygenin HRP.
2. Following TUNEL staining, sections can be immunohistochemically labeled for multiple antigens. *See* Tornusciolo, Schmidt, and Roth, *BioTechniques* 1995; **19**:800–805.
3. When TUNEL labeling cells in tissue culture, decrease the concentration of the dUTP conjugate to 0.10 nmol/100 µL TDT buffer.
4. We have successfully performed TUNEL staining with the following fixatives: 10% formalin, 4% paraformaldehyde, Bouin's solution (at 4°C), Methacarn (60% chloroform, 30% methanol, 10% acetic acid), and Histochoice fixative (Amresco, Solon, OH). Do not use room-temperature Bouin's solution as a fixative, since it will dramatically increase nonspecific nuclear labeling.

Acknowledgments

This work was supported in part by Perkin-Elmer Life Science Products, Inc., and by National Institutes of Health Grant NS35107. I thank Cecelia Latham and Barbara Klocke for expert technical assistance, Angela Schmeckebier for secretarial support, and Dr. Mark Bobrow (Perkin-Elmer Life Science Products) for expert advice and insights on tyramide signal amplification.

References

1. Kuan, C.-Y., Roth, K. A., Flavell, R. A., and Rakic, P. (2000) Mechanism of programmed cell death in the developing brain. *Trends Neurosci.* **23,** 287–293.
2. Clarke, P. G. H. (1990) Developmental cell death: morphological diversity and multiple mechanisms. *Anat. Embryol.* **181,** 195–213.
3. Majno, G. and Joris, I. (1995) Apoptosis, oncosis, and necrosis: an overview of cell death. *Am. J. Pathol.* **146,** 3–15.
4. Kerr, J. F. R., Wyllie, A. H., and Currie, A. R. (1972) Apoptosis: a basic biological phenomenon with wide-ranging implications in tissue kinetics. *Br. J. Cancer* **26,** 239–257.
5. Stadelmann, C. and Lassmann, H. (2000) Detection of apoptosis in tissue sections. *Cell Tissue Res.* **301,** 19–31.

6. Horvitz, H. R. (1999) Genetic control of programmed cell death in the nematode *Caenorhabditis elegans. Cancer Res.* **59,** 1701s–1706s.
7. Korsmeyer, S. J. (1999) BCL-2 gene family and the regulation of programmed cell death. *Cancer Res.* **59,** 1693s–1700s.
8. Nicholson, D. W. (1999) Caspase structure, proteolytic substrates, and function during apoptotic cell death. *Cell Death Differ.* **6,** 1028–1042.
9. D'Sa-Eipper, C., Leonard, J. R., Putcha, G., et al. (2001) DNA damage-induced neural precursor cell apoptosis requires p53 and caspase-9 but neither Bax nor caspase-3. *Development* **128,** 137–146.
10. Yuan, J. and Yankner, B. A. (2000) Apoptosis in the nervous system. *Nature* **407,** 802–809.
11. Sanders, E. J. (1997) Methods for detecting apoptotic cells in tissue. *Histol. Histopathol.* **12,** 1169–1177.
12. Allen, R. T., Hunter, W. J. III, and Agrawal, D. K. (1997) Morphological and biochemical characterization and analysis of apoptosis. *J. Pharmacol. Exp. Ther.* **37,** 215–228.
13. Motoyama, N., Wang, F., Roth, K. A., et al. (1995) Massive cell death of immature hematopoietic cells and neurons in Bcl-x-deficient mice. *Science* **267,** 1506–1510.
14. Willingham, M. C. (1999) Cytochemical methods for the detection of apoptosis. *J. Histochem. Cytochem.* **47,** 1101–1109.
15. Zheng, T. S., Schlosser, S. F., Dao, T., et al. (1998) Caspase-3 controls both cytoplasmic and nuclear events associated with Fas-mediated apoptosis in vivo. *Proc. Natl. Acad. Sci. USA* **95,** 13,618–13,623.
16. Keramaris, E., Stefanis, L., MacLaurin, J., et al. (2000) Involvement of caspase 3 in apoptotic death of cortical neurons evoked by DNA damage. *Mol. Cell. Neurosci.* **15,** 368–379.
17. Kuida, K., Zheng, T. S., Na, S., et al. (1996) Decreased apoptosis in the brain and premature lethality in CPP32-deficient mice. *Nature* **384,** 368–372.
18. Srinivasan, A., Roth, K. A., Sayers, R. O., et al. (1998) In situ immunodetection of activated caspase-3 in apoptotic neurons in the developing nervous system. *Cell Death Differ.* **5,** 1004–1016.
19. Velier, J. J., Ellison, J. A., Kikly, K. K., Spera, P. A., Barone, F. C., and Feuerstein, G. Z. (1999) Caspase-8 and Caspase-3 are expressed by different populations of cortical neurons undergoing delayed cell death after focal stroke in the rat. *J. Neurosci.* **19,** 5932–5941.
20. Fujita, E., Urase, K., Egashira, J., et al. (2000) Detection of caspase-9 activation in the cell death of the Bcl-x-deficient mouse embryo nervous system by cleavage sites-directed antisera. *Dev. Brain Res.* **122,** 135–147.
21. Rohn, T. T., Head, E., Su, J. H., et al. (2001) Correlation between caspase activation and neurofibrillary tangle formation in Alzheimer's disease. *Am. J. Pathol.* **158,** 189–198.
22. Suurmeijer, A. J. H., van der Wijk, J., van Veldhuisen, D. J., Yang, F., and Cole, G. M. (1999) Fractin immunostaining for the detection of apototic cells and apoptotic bodies in formalin-fixed and paraffin-embedded tissue. *Lab. Invest.* **79,** 619–620.
23. Li, X. and Darzynkiewicz, Z. (2000) Cleavage of Poly(ADP-Ribose) polymerase measured in situ in individual cells: Relationship to DNA fragmentation and cell cycle position during apoptosis. *Exp. Cell Res.* **255,** 125–132.

24. Yin, X.-M., Wang, K., Gross, A., Klocke, B., Roth, K. A., and Korsmeyer, S. J. (1999) Bid-deficient mice are resistant to Fas-induced hepatocellular apoptosis. *Nature* **400,** 886–891.
25. Bobrow, M. N., Harris, T. D., Shaughnessy, K. J., and Litt, G. J. (1989) Catalyzed reporter deposition, a novel method of signal amplification: application to membrane immunoassays. *J. Immunol.* **125,** 279–285.
26. Speel, E. J. M., Hopman, A. H. N., and Komminoth, P. (1999) Amplification methods to increase the sensitivity of in situ hybridization: play CARD(S). *J. Histochem. Cytochem.* **47,** 281–288.
27. Shindler, K. S. and Roth, K. A. (1996) Double immunofluorescent staining using two unconjugated primary antisera raised in the same species. *J. Histochem. Cytochem.* **44,** 1331–1335.
28. Zaidi, A. U., Enomoto, H., Milbrandt, J., and Roth, K. A. (2000) Dual fluorescent in situ hybridization and immunohistochemical detection with tyramide signal amplification. *J. Histochem. Cytochem.* **48,** 1369–1376.
29. Collins, J. A., Schandl, C. A., Young, K. K., Vesely, J., and Willingham, M. C. (1997) Major DNA fragmentation is a late event in apoptosis. *J. Histochem. Cytochem.* **45,** 923–934.
30. Adayev, T., Estephan, R., Meserole, S., Mazza, B., Yurkow, E. J., and Banerjee, P. (1998) Externalization of phosphatidylserine may not be an early signal of apoptosis in neuronal cells, but only the phosphatidylserine-displaying apoptotic cells are phagocytosed by microglia. *J. Neurochem.* **71,** 1854–1864.
31. Blankenberg, F. G., Katsikis, P. D., Tait, J. F., et al. (1998) In vivo detection and imaging of phosphatidylserine expression during programmed cell death. *Proc. Natl. Acad. Sci. USA* **95,** 6349–6354.
32. Wyllie, A. H. (1980) Glucocorticoid-induced thymocyte apoptosis is associated with endogenous endonuclease activation. *Nature* **284,** 555–556.
33. Gavrieli, Y., Sherman, Y., and Ben-Sasson, S. A. (1992) Identification of programmed cell death in situ via specific labeling of nuclear DNA fragmentation. *J. Cell Biol.* **119,** 493–501.
34. Naruse, I., Keino, H., and Kawarada, Y. (1994) Antibody against single-stranded DNA detects both programmed cell death and drug-induced apoptosis. *Histochemistry* **101,** 73–78.
35. Naruse, I. and Keino, H. (1995) Apoptosis in the developing CNS. *Prog. Neurobiol.* **47,** 135–155.
36. Frankfurt, O. S. and Krishan, A. (2001) Identification of apoptotic cells by formamide-induced DNA denaturation in condensed chromatin. *J. Histochem. Cytochem.* **49,** 369–378.
37. McCarthy, N. J. and Evan, G. I. (1998) Methods for detecting and quantifying apoptosis. *Curr. Topics Dev. Biol.* **36,** 259–278.
38. Ishimaru, M. J., Ikonomidou, C., Tenkova, T. I., Der, T. C., Dikranian, K., Sesma, M. A., and Olney, J. W. (1999) Distinguishing excitotoxic from apoptotic neurodegeneration in the developing rat brain. *J. Comp. Neurol.* **408,** 461–476.
39. de Torres, C., Munell, F., Ferrer, I., Reventós, J., and Macaya, A. (1997) Identification of necrotic cell death by the TUNEL assay in the hypoxic-ischemic neonatal rat brain. *Neurosci. Lett.* **230,** 1–4.
40. Labat-Moleur, F., Guillermet, C., Lormier, P., et al. (1998) TUNEL apoptotic cell detection in tissue sections: critical evaluation and improvement. *J. Histochem. Cytochem.* **46,** 327–334.

41. Tornusciolo, D. R. Z., Schmidt, R. E., and Roth, K. A. (1995) Simultaneous detection of TDT-mediated dUTP-biotin nick end-labeling (TUNEL)-positive cells and multiple immunohistochemical markers in single tissue sections. *BioTechniques* **19,** 2–7.
42. Didenko, V. V., Tunstead, J. R., and Hornsby, P. J. (1998) Biotin-labeled hairpin oligonucleotides. Probes to detect double-strand breaks in DNA in apoptotic cells. *Am. J. Pathol.* **152,** 897–902.
43. Zhu, C., Wang, X., Hagberg, H., and Blomgren, K. (2000) Correlation between caspase-3 activation and three different markers of DNA damage in neonatal cerebral hypoxia-ischemia. *J. Neurochem.* **75,** 819–829.
44. Slater, M. and Murphy, C. R. (1999) Detection of apoptotic DNA damage in prostate hyperplasia using tyramide-amplified avidin-HRP. *Histochem. J.* **31,** 747–749.

Apoptotic and Oxidative Indicators in Alzheimer's Disease

*Arun K. Raina, Lawrence M. Sayre,
Craig S. Atwood, Catherine A. Rottkamp,
Ayala Hochman, Xiongwei Zhu,
Mark E. Obrenovich, Shun Shimohama,
Akihiko Nunomura, Atsushi Takeda,
George Perry, and Mark A. Smith*

Introduction

Dogma suggests that cell death mechanisms can present with either a necrotic or an apoptotic phenotype. Recent evidence, however, seems to point to a far more complex picture, in which apoptotic and necrotic phenotypes might present simultaneously (1). For example, TUNEL positivity in cardiomyocytes does not necessarily entail the presence of cell death that has an apoptotic phenotype (2). Complicating the matter further is the promiscuous use of the same term, apoptosis, to mean different things at different times, by conflating process (a cell death program) with product (the apoptotic phenotype). Cell death can be classified into *programmed cell death*, which entails a global/extrinsic program of cell death, and a *cell death program(s)*, which entails a local/intrinsic/ cellular death program (3). The latter can present in a variety of phenotypes ranging from necrosis to apoptosis or as a combination phenotype (1), while the former is seen primarily during development and presents with an apoptotic phenotype. These ideas are useful when we come across novel phenomena or in situations where there is ambiguity as to the nature of cell death, i.e., Alzheimer's disease (AD), where the earliest perceptible event is the presence of oxidative stress (4,5). Indeed, the presence of oxidative

From: *Neuromethods, Vol. 37: Apoptosis Techniques and Protocols, 2nd Ed.*
Edited by: A. C. LeBlanc @ Humana Press, Inc., Totowa, NJ

stress markers in AD parallels neuronal susceptibility to cell death in AD *(4,6,7)*. Here we will review methods to detect both the proximal event (oxidative stress) as well as the most distal event (cell death) in AD.

2. Mechanisms of Cell Death in Alzheimer's Disease

A great deal of the research in AD over the last several decades has concentrated on the precise mechanisms responsible for the death of susceptible neurons. This is complicated by the fact that the nature and time-course of neuronal cell death in AD is controversial *(8,9)*. The two cellular events that are a feature in AD include oxidative stress and apoptosis, with oxidative stress preceding apoptosis. As evidence supporting an apoptotic mechanism was continually increasing, questions began to be asked about the plausibility of this claim. Although various studies point to neuronal death in AD as being the result of a cell death program that results in an apoptotic phenotype, the stereotypical manifestations that characterize the terminal phases of this apoptotic cell death program, such as detachment, chromatin condensation, nuclear segmentation, blebbing, and apoptotic bodies, are not seen either together or by themselves in AD and not for lack of trying. Furthermore, apoptosis requires only 16–24 h for completion and therefore, in a chronic disease like AD with an average duration of almost 10 yr, less than one in about 4000 cells should be undergoing apoptosis at any given time; i.e., observation of apoptotic events should be rare *(10,11)*. Indeed, if all the neurons that are reported as having DNA cleavage were undergoing apoptosis, the brain would rapidly be stripped of neurons and the symptomatology would be abbreviated to months and not years—this is certainly not the case in AD.

A vast array of putatively apoptogenic stimuli present in neurons in AD can act alone and/or synergistically with reactive oxygen species (ROS) *(12)*, amyloid-β *(13)*, energy failure *(14)*, and HNE oxidants *(7,15)*. Added to this is the apoptotic risk posed in familial AD *(16)*. However, although internucleosomal DNA fragmentation and end-labeling techniques have been used as absolute indices of apoptotic cell death in AD *(17,18)*, it is now apparent that DNA fragmentation, in addition to being inferentially related to apoptosis, is also seen secondary to oxidative stress *(6,19,20)* and postmortem autolysis *(21)*. Indeed, unlike DNA fragmentation, the hallmark end-stage signs of an apoptotic cell death program such as nuclear chromatin condensation and apoptotic bodies are not seen in AD *(10)*. Added to this mechanistic and biochemical ambiguity of neuronal death mechanisms in AD is the already-mentioned temporal dichotomy between the acuteness of an apoptotic cell death program and the chronicity of AD *(11)*. Finally, it is puzzling as to why we plainly can locate mitotic substages in most tissues and yet fail to see substantive evidence of a cell death program with an apoptotic phenotype, given that in an adult environment, mitosis takes only about an hour to complete.

In light of these arguments, it becomes important to revisit the nature of neuronal cell death in AD, by considering the methodology used in determining cell death. Given the lack of adherence to standards, it is easy to comprehend how contradictory reports regarding the nature of cell death in novel scenarios can arise.

Methods of detecting apoptotic mechanisms have been developed that assay for markers all along the apoptotic pathway, including the presence of death receptors, changes in the plasma membrane, activation of the caspase protease cascade, changes in mitochondrial integrity, and fragmentation of nuclear DNA. However, for the purpose of this discussion, it is most convenient to separate all of these methods into two categories, those that detect changes in cell populations and those that detect changes in individual cells. Each of these procedures has its own sets of advantages and disadvantages, but what remains consistent is that no single criterion can be used to definitively establish the presence of apoptotic processes. Only when evidence from several of the currently available methods is pooled is it possible to conclude either the presence or absence of apoptotic cell death.

3. Apoptosis Detection
Based on Morphological Criteria

The first established criterion for cell death via an apoptotic mechanism was the characteristic cell morphology including blebbing and nuclear condensation and segmentation. The nuclear features of the apoptotic process that have been identified include condensation of chromatin around the periphery of the nucleus and fragmentation of the nucleus into multiple chromatin bodies surrounded by nuclear envelope remnants. Ironically, enucleation does not prevent apoptosis from going forward in the cytoplasm.

These changes, which can be detected by either light or electron microscopy, are the most accurate indicators for the involvement of apoptosis in the death of any given cell. This is true in spite of the fact that the exact time-course of these nuclear characteristics remains unclear and that these changes do not seem necessary for the completion of apoptosis given that enucleated cells can complete the process *(22)*. The fact that such morphological changes are considered the gold standard for the presence of apoptosis calls into question the claim that cell death in AD is apoptotic given that these hallmark end-stage signs of an apoptotic cell death program such as nuclear chromatin condensation and apoptotic bodies are not seen *(10)*.

Electron microscopy conclusively reveals the presence of apoptosis, although it cannot quantify this since apoptosis is in most cases asynchronous. Other microscopic techniques involve fluorescent microscopy, after staining with propidium iodide or 4,6-diamidino-2-phenolindole, while labeled annexin-V *(23)* can show phosphatidyl serine externalization. Finally, light microscopy can be used during hematoxylin and eosin staining, which involves primarily pattern recognition by a neuropathologist of the pyknotic and hyperchromatic nucleus. Nuclear segmentation can also be seen using this inexpensive technique.

4. Apoptosis Detection Based on DNA Fragmentation

Apoptosis detection using DNA fragmentation is among the most widely used criteria for detection of apoptosis based on the fact that in apoptosis cation-dependent endonucleases cleave genomic DNA between nucleosomes. DNA fragmentation laddering depends on the fact that, in the case of apoptotic cell death, one expects to see mono- and oligonucleosomal-sized DNA fragments as opposed to the high-molecular weight-species normally seen with genomic DNA, creating a distinctive electrophoretic banding pattern in agrose gel systems. While this method can be used effectively when evaluating cells in vitro and in biopsy tissue, its value in establishing apoptotic processes in AD is limited. In fact, assays based on DNA fragmentation may represent the greatest source of misunderstanding in associating AD with apoptosis, because similar mono- and oligonucleosomal patterns can be seen in a variety of conditions in which histones protect DNA from a variety of insults, including necrosis and oxidative damage. Another confounding issue in the use of DNA laddering is that this type of alteration can also be characteristic of postmortem autolysis *(21)*, a problem that is unavoidable when studying neuronal death in AD.

In situ DNA fragmentation can also be detected by end-labeling techniques that detect the degradation of DNA by taking advantage of DNA repair enzymes to add labeled nucleotides, either nicked strands according to the opposite strand template or addition of nucleotides independent of a template, either at nick sites or the blunt-end double-strand breaks. Respectively, these techniques are called in *situ* nick translation (ISNT) and TUNEL (TdT-mediated X-UTP nick end labeling). Of these techniques, TUNEL is currently the most commonly used. The principle behind terminal end labeling is that the enzyme terminal deoxyribonucleotidyl transferase (TdT) will catalyze the addition of nucleosides at free 3' hydroxyl groups produced by apoptotic endonucleases. Under the appropriate conditions, including the divalent cation Co^{2+}, the TdT will add nucleotides nonspecifically to blunt 3' ends, 3' protruding ends, or a 3' end that is recessed within a DNA strand *(24)*. In the case of *in situ* TUNEL, tissue sections are fixed and permeabilized and then treated with TdT in the presence of deoxyuridine coupled to a chromogenic or fluorescent label that is then detected by light microscopy *(25)*.

The advantage of these techniques is that it allows the investigator to determine the number and specific populations of cells that are undergoing an apoptotic process at any given time and determine the time-course of DNA fragmentation and other nuclear morphological changes. Although this has proven to be a powerful technique in understanding the process of apoptosis, it must be used with a clear understanding of its limitations and confounding complications. First and foremost, the results of TUNEL assays are highly dependent on the manner in which the tissue is preserved and prepared *(26)*. This is of great importance in the study of AD tissue, which often is not fixed until many hours postmortem. Additionally, nuclear DNA fragmentation caused by other modes of cell death can lead to false positive results in a large number of neurons in a given population *(27)*, while false negative results can be obtained if the fragmentation is not accessible to the TdT enzyme after permeabilization *(28)*. The former has led some investigators to pretreat sections with proteases so as to expose inaccessible sites, which has unknown consequences on the TUNEL results *(29)*.

5. Detection of Apoptosis
Based on Effectors of Cytoplasmic Changes

Despite the extremely highly conserved nature that has been described for the apoptotic process, the upstream effectors of this mode of cell death have proven highly variable. One aspect, however, that has proven to be fairly conserved is the activation of a cascade of proteases of the caspase family, which are thought to mediate the very early stages of apoptosis. Although caspases are known to cleave a large number of cellular substrates after aspartate residues, their main action in apoptosis is to cleave other caspase family members. All of the caspases are synthesized as proenzymes that are activated by autocleavage mediated by one of a number of extracellular stimuli or by cleavage performed by another caspase family member that lies upstream of it. Caspases are generally classified into one of two groups, the initiator (caspases 8 and 9) as well as the executioner proteins (caspases 3, 6, and 7) of apoptosis. The initiator caspases are those that lie directly downstream of extracellular or intracellular activators, while the activation of the executioner capases are thought to lead directly to the commencement of the apoptotic cell death program. Due to their role as the most important effectors of apoptosis, a great deal of effort has been focused toward using these proteases to detect apoptosis in both cells and cell populations. Also, because activation of caspases is thought to be the earliest known stage in the apoptotic process, detection of caspase activation is often favored over characterization of nuclear and cytoplasmic phenotypic changes.

The effector of the proximal caspase machinery includes the cleavage of the 116-kDa poly (ADP-ribose) polymerase (PARP) molecule into an 85-kDa fragment. PARP is thus prevented from entry into the nucleus and affecting its DNA repair activity. Other effects of the caspase amplification system include removal of the inhibitor subunit from inhibitor of caspase-activated deoxyribonuclease to yield caspase-activated deoxyribonuclease, which is involved in DNA fragmentation, lamanin A, actin, Gas 2, and fodrin, each of which can be detected.

The classic way to assay for the presence of activated caspases has been to incubate homogenates with known substrates of these proteases and subsequently to detect cleavage products of these same substrates. The most significant problem with this type of assay is that caspases tend to be rather promiscuous proteases, capable of cleaving a large number of substrates on the C-terminal side of the aspartate residues. Also, many different caspase members are capable of cleaving any given substrate at the same residues.

Our laboratory saw the immunocytochemical detection of caspases as an important tool that could be used to clarify some of the confounding issues plaguing our understanding of apoptotic processes in AD. We undertook a systematic study of the caspase cascade proteins in AD by evaluating the key initiator (caspases 8 and 9) as well as the executioner proteins (caspases 3, 6, and 7) of apoptosis *(30)*.

Our results show that while upstream caspases are present in association with the pathological lesions in all cases of AD, downstream caspases including 3, 6, and 7 that signal completion of the apoptosis pathway remain unchanged in such neurons. This lack of downstream amplification of signaling via the caspase pathway may well account for the lack of an apoptotic phenotype in AD. Furthermore, the neurons that are present and viable at latter stages in AD must have employed survival compensations to the presumptive proapoptotic environment encountered in AD *(30)*.

The presence of caspase-8, a downstream component of the Fas/FADD pathway and a specific activator of caspase-3 *(31)*, along with caspase-9, which through the downstream mitochondrial pathway also leads to caspase-3 activation, argues for the recruitment of the initiator components of the caspase pathway in neurons of AD. However, the lack of increased caspase-3 and 7 in the neuropathology of AD and the sporadic incidence of other crucial downstream events of the caspase cascade *(32)* indicates incomplete or effectively absent amplification of the upstream apoptotic signal. Indeed, since such downstream caspases and their proteolytic products are recognized as markers of apoptotic irreversibility *(33)*, their avoidance or sporadic appearance in AD indicates an absence of effective distal propagation of the caspase-mediated apoptotic signal(s). Indeed, caspase 3 and 9 activity are important proapoptotic regulators of postmitotic neuronal homeostasis *(34)*.

Recently, activated caspase-3 has been shown to be present within autophagic granules and rarely within neurons in AD and Down syndrome cases *(35,36)* as well as in association with senile plaques *(37)*. However, the localization of caspase-3 activity, like that of caspase-6 in our study *(30)*, within a select subcellular compartment, i.e., autophagic granules *(36)* or the extracellular compartment *(37)*, does not necessarily indicate a global, effective caspase amplification. Moreover, modulation of distal substrates of activated caspase-3 may lead to further modification of this cell death pathway and may explain the lack of an apoptotic death pathway. This may explain the lack of evidence demonstrating the acute end stages, including nuclear compensation and blebbing, in AD-susceptible neurons *(10,11)*.

Apoptosis can be prevented by antiapoptotic members of the Bcl-2 family, many of which have been reported in AD, including Bcl-xl and Bcl-w (reviewed in ref. *38*; Zhu et al., unpublished data). Interactions of activated caspases with XIAPs as well as SMAC (DIABLO) provide for postactivation regulation of the death signal *(39)*. Additional evidence for the antiapoptotic nature of the intraneuronal environment in AD includes the expression of GADD 45, a growth-arrest DNA-damage-inducible protein in select neurons in AD, where its early expression is associated with the expression of Bcl-2. Thus, this may in turn confer survival advantages in AD *(40)*. Furthermore, the hyperphosphorylated intraneuronal state that exists in AD acts to inhibit downstream substrate proteolysis and thus can promote neuronal survival. Indeed, accumulating evidence points to dephosphorylation being associated with increased capability to cleave PARP *(41)*. Additionally, chronic oxidative stress, a major component of early AD pathophysiology *(42)*, would further inhibit downstream propagation of caspase-mediated apoptotic signals *(43)*.

The upregulation of individual caspases, while seemingly not leading to apoptosis, could have great significance related to the pathogenesis of AD. For example, caspase-6, found in association with amyloid-β senile plaques *(30)*, cleaves amyloid-β protein precursor (AβPP), resulting in a 6.5-kDa amyloid fragment, and is proposed as an alternative AβPP processing pathway *(44)*. The lack of caspase-6, a downstream effector caspase, in neuronal pathology, however, argues against a specific role in the apoptotic cascade.

Given that execution of apoptosis requires amplification of the caspase-mediated apoptotic signal, our results indicate that, in AD, there is a lack of effective apoptotic signal propagation to downstream caspase effectors. AD represents that first in-vivo situation reported in which the initiation of apoptosis does not lead directly to apoptotic cell death, an interpretation consistent with earlier reports *(45–49)*. Obviously, in certain brain areas in AD, while many neurons do degenerate, it is unclear whether this is through successful propogation of the apoptotic cascade or by another pathway such as paratosis *(1)*. However, in those surviving neurons, it is clear that neuronal viability in AD is, in part, maintained by the lack of distal transmission of the caspase-mediated apoptotic signal(s). This novel phenomena, which we term *abortosis (30)*, leads to apoptotic avoidance and hence selects for neuronal survival rather than a caspase-induced cell death program that presents as apoptosis.

6. Neurons are Vulnerable to Oxidative Stress

Neurons of the central nervous system are subject to a number of unique conditions that make them particularly vulnerable to oxidative stress and its sequelae that can culminate in cell death. This vulnerability is a consequence of the excessive oxygen utilization, the high unsaturated lipid content of neuronal membranes as compared to normal plasma membrane, and the postmitotic nature of primary neurons. Due to their unique status, a balance between the production of ROS and antioxidant defenses is of particularly significance in neurons. Under normal physiological conditions, damage produced by oxygen radicals is kept in check by an efficient array of antioxidant systems that display an impressive level of redundancy, which provide multiple lines of defense. However, in cases of age-related neurodegeneration, like that observed in AD, this balance between oxidative radicals and antioxidant defenses is altered, allowing various forms of cellular and molecular damage. This disequilibrium is intensified by the simultaneous low levels of the major neuronal antioxidant, glutathione *(50)*. Additionally, neurofilaments and cytoskeletal proteins such as tau, which are particularly abundant in neurons, have relatively long half-lives and contain high proportions of lysine residues. The presence of lysine residues makes these proteins particularly susceptible to modification by carbonyl-containing products of glycoxidation and lipoxidation. Such modifications or direct oxidation results in neutralization of the charge and consequent alteration of protein–protein electrostatic interactions. The latter provides a key *noncovalent* mechanism for aberrant protein aggregation/deposition even in the absence of covalent crosslinking.

Oxidative stress appears to be a common contributor to both pathological protein deposition and irreversible cellular dysfunction leading to neuronal death, although there is a complex interplay of biochemical mechanisms in force in neurodegenerative disease that suggests several possible investigative avenues. By taking advantage of the increasing availability of immunochemical tools that recognize specific oxidative markers on pathological deposits, the question of whether oxidative stress is contributory to, or merely a consequence of, the long-term presence of abnormal protein deposition characteristic of AD, including neurofibrillary tangles (NFT) and senile plaques, may very well be answerable on the basis of spatio-temporal patterns of epitope appearance. Although the various neurodegenerative diseases are connected by the presence of cytoskeletal abnormalities, it is unlikely that the vulnerable neurons (pyramidal neurons in the hippocampus/cortex in AD, pigmented neurons in the substantia nigra in Parkinson disease, and motor neurons in amyotrophic lateral sclerosis) respond identically to oxidative stresses. Thus, while a certain degree of overlap among these diseases in associated oxidative stress biochemistry might be expected, each individual disease will likely show a distinct oxidative profile and therefore no *individual* marker should be relied upon as diagnostic of oxidative stress.

Here we expand upon *in situ* detection methods that can be used for qualitative and semiquantitative measurement of various oxidative stress markers in tissue sections obtained postmortem. Though the presence of oxidative stress damage implies cytotoxicity, it is not our intention to suggest that the presence of particular markers correlates with a specific mechanism of toxicity. Any such correlation requires a consideration of whether the damage represents a primary or secondary event in the cascade of biochemical processes leading to neuronal death.

7. The Role of *In Situ* Methodology

The value of using *in situ* methods rather than bulk analysis of tissue samples cannot be overemphasized when considering a system as structurally and cellularly complex as the nervous system. In addition to the fact that vulnerable neurons comprise only a small percentage of the affected brain region, it is essential to consider other cell types that may be compromised by the disease state, such as glia. It is quite possible that, while neurons may display one type of damage or response, other support cells may demonstrate compensatory responses disparate from those displayed by neurons. While at present this point is far from being established, it is enough of a theoretical concern to undertake consistent *in situ* studies in parallel with bulk analysis experiments. Further, the results of bulk analysis of oxidative damage are complicated by the contribution of long-lived proteins that are modified constantly by the process of normal physiological aging. Therefore, the same properties that make neurofilaments and other cytoskeletal proteins ideal monitors for the oxidative changes associated with aging limits their sensitivity to detect disease-specific changes. While isolation of neurons or other cell fractions may increase relative specificity, this is not generally sufficient in brain tissue, where there is always the concern of vascular contamination. Therefore, the information acquired using *in situ* is very different qualitatively from that which is attainable using bulk analysis methods and provides valuable data regarding the spatio-temporal aspects of disease pathogenesis.

8. The Biochemistry of Oxidative Stress

Direct protein oxidation, mediated by diffusible hydroxyl radical ($\bullet$OH) or metal-catalyzed "site-specific" oxidation at the sites of metal-ion binding *(51,52)*, are the most obvious form of protein damage resulting from oxidative imbalance. This type of interaction can result in backbone cleavage events or in the specific oxidation of side-chain moieties, sometimes involving crosslinking reactions. Since oxidative stress is associated with high local concentrations of both superoxide and nitric oxide, produced by the inducible isoform of nitric oxide synthase, the product of their combination, peroxynitrite, has become an important secondary oxidative stress marker. The formation of peroxynitrite can then lead to the nitration and/or oxidation of Tyr and Trp residues *(53,54)*.

The large majority of oxidative-stress-dependent protein modification, however, involves the adduction of products following the oxidation of carbohydrate or lipid moieties (glycoxidation and lipoxidation, respectively). Modification of protein-based lysines and arginines by reducing sugars initiates a complex sequence of oxidation, rearrangement, condensation, and fragmentation reactions, known collectively as the Maillard reaction *(55)*, which ultimately results in the formation of stable adducts called advanced glycation end products (AGEs). The modification of proteins by lipoxidation products, such as 4-hydroxy-2-nonenal (HNE) and malondialdehyde (MDA), form adducts that are initially unstable but subsequently evolve over time to give stable advanced lipoxidation end products (ALEs) *(56,57)*. While it is widely recognized that protein modification and the formation of carbohydrate and lipid adduction products can be of great importance individually, substantial evidence now suggests that each of these mechanisms can synergistically promote the others, accelerating damage due to oxidative imbalance. It is hypothesized that this may be the result of condensation of reducing sugars with amino groups that form moieties capable of redox cycling *(58)*.

9. Markers of Oxidative Stress

9.1. Detection of Peroxynitrite Damage

The complex chemistry of peroxynitrite has been the subject of intense recent study, and it has become clear that there are two principal pathways for protein modification: one involving homolytic hydroxyl radical-like chemistry that results in protein-based carbonyls, and the other involving electrophilic nitration of vulnerable electron-rich aromatic ring side chains, particularly those of Tyr and Trp *(53,54)*. The activity of peroxynitrite can be augmented by forming a CO_2 adduct, which augments its reactivity when in the presence of bicarbonate. Formation of 3-nitrotyrosine by this route has become the classical protein marker for the presence specifically of peroxynitrite *(6)*. Affinity-purified rabbit antiserum or mouse monoclonal antibodies (7A2 or 1A6), raised to nitrated keyhole limpet hemocyanin *(59)*, can be used to detect nitrotyrosine modifications immunohistochemically.

9.2. Advanced Glycation
and Lipoxidation End Products (AGEs and ALEs)

The glycation of proteins by reducing sugars (Maillard reaction) is a key posttranslational event and can result in time-dependent adduct formation that becomes especially susceptible to oxidative elaboration. In addition, oxidative stress can result in oxidized sugar derivatives that can subsequently modify proteins through a process known as glycoxidation. Similarly, and of more general importance, oxidative stress results in lipid peroxidation and the production of a range of aldehydes capable of modifying numerous proteins. Although there are a number of chemical events that occur early and rapidly, these are of less pathophysiological importance due to their reversible nature. Therefore, identifying more stable proteins that accumulate over time, such as AGEs and ALEs, has proved much more efficacious. Protein modification can result in both noncrosslink and crosslink AGEs and ALEs, the latter arising from the potential bifunctional reactivity, such as of the lipid-derived modifiers HNE and malondialdehyde. Examples of crosslink adducts that can be detected are fluorescent Arg-Lys AGE crosslink pentosidine *(60)*, fluorescent Lys-Lys ALE crosslink pyrrolidinone iminium *(61)*, fluorescent 3-(1,4-dihydropyridin-4-yl) pyridinium, 1-amino-3-iminopropene Lys-Lys crosslinks arising from MDA *(57)*, and the imidizolium Lys-Lys crosslinks GOLD and MOLD arising from glyoxal and methylglyoxal, respectively *(62)*. These modifications are associated with the chemistry of both AGEs and ALEs. Examples of noncrosslink modifications are lysine-derived AGE pyrraline *(63)*, HNE-derived 2-pentylpyrrole *(56,64)*, carboxymethyllysine (CML) *(65)*, resulting from glyoxal or from fragmentation of a glycated precursor, and the fluorescent arginine-based argpyrimidine arising from methylglyoxal *(66)*.

We and others *(67–69)* have determined that AGE modifications are present on both NFT and senile plaques in AD. These studies employed antibodies not only to the specific structures pentosidine and pyrraline, but also less defined AGE antibodies raised to carrier proteins treated with reducing sugars for long periods of time. AGE modification is likely an early *in-vivo* event, since both diffuse senile plaques *(67)* and paired helical filaments of τ *(68,69)*, considered two of the earliest pathological changes in AD *(70,71)*, are specifically recognized by anti-AGE antibodies. The sensitivity of detection of protein-based carbonyls can be optimized by derivitizing them with 2,4-dinitrophenylhydrazine. This permitted the *in situ* detection of carbonyl reactivity not only within NFT but also within vulnerable neurons in AD *(42,72)*.

Immunostaining using antibodies to lipoxidation-derived modifications has exhibited more distinctive patterns. Antibodies to characterized HNE adducts are localized exclusively to neuronal cell bodies and neurofibrillary pathology *(7,73)*. The fact that antibodies to the HNE-lysine-derived pyrrole, an ALE, stain not only intraneuronal and extraneuronal NFT, but also apparently normal hippocampal neurons, points to the early role that oxidative processes play in AD pathogenesis. This same profile of staining is seen using antibodies to markers of direct protein oxidation, including nitrotyrosine and protein-based carbonyls *(6,42,72)*.

9.3. Nucleic Acid Damage

Examination of the spatiotemporal relationship between the presence of these classical markers of oxidative stress and the hallmark lesions of AD revealed an apparent paradox. Namely, we found that while stable alterations such as glycation products are associated predominantly with NFT and senile plaques *(67)*, whereas reversible or rapidly degraded adduction products are largely restricted to the cytoplasmic compartment of susceptible neurons. This indicates that oxidative damage is not limited to prolonged lesions and represents one of the earliest detectable events in the cellular disease course of AD. Therefore, like previously discussed, the crosslink modification of proteins, which slow their turnover, do not accurately assess the recent state of affairs in the tissue of interest, but instead provide an assessment tool for long-term history of disease. Another avenue to explore is oxidative modification of nucleic acids, which are cellular components that can be altered in a number of ways including base modifications, substitutions, and deletions. This can involve direct oxidation (such as the addition of •OH to purine and pyrimidine rings, and the •OH-mediated cleavage of the sugar backbone) or adduction of lipoxidation products such as MDA and HNE.

The key to specifying the spatiotemporal aspects of the oxidative imbalance in AD was resolved using a specific antibody to 8-hydroxyguanosine (8OHG), which is a modification of nucleic acids derived largely from hydroxide (•OH) attack of guanidine. By examining the modification of cytoplasmic RNA, which is by definition a short-lived molecule, it is therefore possible to develop a clearer picture of the present state of the cell. Our studies of this marker in AD revealed that not only is 8OHG-modified RNA greatly increased in the cytoplasm of vulnerable neurons, but also that its presence is inversely correlated with the deposition of amyloid-β plaques and the presence of NFT *(4)*. This has allowed for a new line of thought to develop, which maintains that the hallmark lesions associated with AD may reflect the presence of a protective mechanism in addition to, or in fact in opposition to, a pathogenic role.

9.4. Cellular Response Factors

Cells are not passive to increased oxygen radical production but rather upregulate protective antioxidant defenses in response to oxidative stress. It is the failure to do so effectively that leads to cellular and molecular damage that can eventually lead to irreversible mitochondrial damage and cell death. In neurodegenerative diseases, heme oxygenase-1 (HO-1), an enzyme that converts heme, a pro-oxidant, to biliverdin/bilirubin (antioxidants) and free iron and induction is coincident with formation of NFT *(74)*. The overexpression of HO-1 has been established as among the most sensitive and selective indicators of cellular oxidative response *(75)*. Both HO-1 and its mRNA have been shown to be increased in the brains of AD patients when compared to age-matched controls *(67,74,76)*. Furthermore, quantitative immunocytochemistry of AD brains sections shows a complete overlap between those neurons overexpressing HO-1 and those that contain oxidatively modified τ *(77,78)*. Seen in the context of arresting apoptosis, HO-1 and τ may coordinately play a role in maintaining the neurons free from the apoptotic signal cytochrome-*c*, since τ has strong iron-binding sites *(79)*.

Activation of stress-induced signal transduction pathways, such as p38 and JNK/SAPK, is also a cardinal feature of stressed cells. In this regard, it is notable that both of these pathways are activated in neurons in AD *(80,81)*. Additionally, since p38 is also a cell cycle-regulated phenomena, such findings provide crucial a link between the triad of cell cycle reentrant phenotype, cytoskeletal phosphorylation, and oxidative stress in AD (82).

9.5. Redox Metal Imbalance

Free iron, more than any other transition metal, has been implicated in undergoing redox transitions in vivo with the consequent generation of oxygen free radicals. Given the importance of iron as a catalyst for the generation of ROS, changes in proteins associated with iron homeostasis can be used as an index of a cellular response. One such class of proteins, the iron regulatory proteins (IRP), responds to cellular iron concentrations by regulating the translation of proteins involved in iron uptake, storage, and utilization *(83,84)*.

Additionally, several studies have implicated imbalances of trace metals in AD, including aluminum, silicon, lead, mercury, zinc, copper, and iron. Zinc(II), iron(III), and copper(II) are also significantly increased in AD neuropil, and these metals are further concentrated within the core and periphery and senile plaques *(85)*. *In situ* methods for detection of iron(II) and iron(III) demonstrated a significant association with both NFT and senile plaques in AD *(86)*, which may be due partly to the ability of iron to bind to their primary protein constituents, τ and amyloid-β, respectively *(87,88)*. We recently reported that the direct detection of redox active iron in AD lesions is prevented by first exposing tissue sections to iron-selective chelators and can be reinstated by reexposure to iron salts *(89)*. Therefore, it is probable that the accumulation of redox active metals is a major source of ROS in AD and that the detection of these metals represents a valid measure of oxidative stress in vivo.

10. Conclusion

This brief overview highlights many of the recent developments in the area of oxidative stress and apoptosis with special reference to AD. Overall, the panel of *in situ* methods for localizing apoptosis and oxidative stress damage in AD is permitting an assessment of the spatio-temporal aspects to neuronal degeneration that define the disease process. One question that will ultimately be answered through better characterization of the cellular response to AD lesions is why certain oxidative stress markers show up only in the neurofibrillary pathology of neurons, whereas others are associated in vulnerable neurons without pathology.

The biochemical cell death pathways that lead to the morphotype, which we call apoptosis ("falling off"), are many. As of yet, there is no single type of evidence that by itself will form the sufficient and necessary criteria that lead to the conclusion of cell death by a particular apoptotic mechanism. However, if we do not conflate mechanism with end-stage histopathological features, we think that confusion can be diminished as well as novel environments such as AD better understood, where our evidence argues for neurons trying to avoid apoptosis—a process that we have termed *abortosis*.

References

1. Sperandio, S., de Belle, I., and Bredesen, D. E. (2000) An alternative, non-apoptotic form of programmed cell death. *Proc. Natl. Acad. Sci. USA* **97,** 14,376–14,381.

2. de Boer, R. A., van Veldhuisen, D. J., van der Wijk, J., et al. (2000) Additional use of immunostaining for active caspase 3 and cleaved actin and PARP fragments to detect apoptosis in patients with chronic heart failure. *J. Card. Fail* **6,** 330–337.

3. Ratel, D., Boisseau, S., Nasser, V., Berger, F., and Wion, D. (2001) Programmed cell death or cell death programme? That is the question. *J. Theor. Biol.* **208,** 385–386.

4. Nunomura, A., Perry, G., Pappolla, M. A., et al. (1999) RNA oxidation is a prominent feature of vulnerable neurons in Alzheimer's disease. *J. Neurosci.* **19,** 1959–1964.

5. Nunomura, A., Perry, G., Aliev, G., et al. (2001) Oxidative damage is the earliest event in Alzheimer's disease. *J. Neuropathol. Exp. Neurol.* **60,** 759–767.

6. Smith, M. A., Harris, P. L. R., Sayre, L. M., Beckman, J. S., and Perry, G. (1997) Widespread peroxynitrite-mediated damage in Alzheimer's disease. *J. Neurosci.* **17,** 2653–2657.

7. Sayre, L. M., Zelasko, D. A., Harris, P. L. R., Perry, G., Salomon, R. G., and Smith, M. A. (1997) 4-Hydroxynonenal-derived advanced lipid peroxidation end products are increased in Alzheimer's disease. *J. Neurochem.* **68,** 2092–2097.

8. Bancher, C., Lassmann, H., Breitschopf, H., and Jellinger, K. A. (1997) Mechanisms of cell death in Alzheimer's disease. *J. Neural Transm. Suppl.* **50,** 141–152.

9. Jellinger, K. A. and Bancher, C. (1998) Neuropathology of Alzheimer's disease: a critical update. *J. Neural Transm. Suppl.* **54,** 77–95.

10. Perry, G., Nunomura, A., Lucassen, P., Lassmann, H., and Smith, M. A. (1998) Apoptosis and Alzheimer's disease (Letter). *Science* **282,** 1268–1269.

11. Perry, G., Nunomura, A., and Smith, M. A. (1998) A suicide note from Alzheimer disease neurons? (News and Views). *Nature Med.* **4,** 897–898.

12. Slater, A. F., Stefan, C., Nobel, I., van den Dobbelsteen, D. J., and Orrenius, S. (1995) Signalling mechanisms and oxidative stress in apoptosis. *Toxicol. Lett.* **82–83,** 149–153.

13. Yankner, B. A. (1996) New clues to Alzheimer's disease: unraveling the roles of amyloid and tau. *Nature Med.* **2,** 850–852.

14. Vander Heiden, M. G., Chandel, N. S., Li, X. X., Schumacker, P. T., Colombini, M., and Thompson, C. B. (2000) Outer mitochondrial membrane permeability can regulate coupled respiration and cell survival. *Proc. Natl. Acad. Sci. USA* **97,** 4666–4671.

15. Mark, R. J., Lovell, M. A., Markesbery, W. R., Uchida, K., and Mattson, M. P. (1997) A role for 4-hydroxynonenal, an aldehydic product of lipid peroxidation, in disruption of ion homeostasis and neuronal death induced by amyloid beta-peptide. *J. Neurochem.* **68,** 255–264.

16. Deng, G., Pike, C. J., and Cotman, C. W. (1996) Alzheimer-associated presenilin-2 confers increased sensitivity to apoptosis in PC12 cells. *FEBS Lett.* **397,** 50–54.

17. Anderson, A. J., Su, J. H., and Cotman, C. W. (1996) DNA damage and apoptosis in Alzheimer's disease: colocalization with c-Jun immunoreactivity, relationship to brain area, and effect of postmortem delay. *J. Neurosci.* **16,** 1710–1719.

18. Cotman, C. W. and Su, J. H. (1996) Mechanisms of neuronal death in Alzheimer's disease. *Brain Pathol.* **6,** 493–506.

19. Tsang, S. Y., Tam, S. C., Bremner, I., and Burkitt, M. J. (1996) Research communication copper-1,10-phenanthroline induces internucleosomal DNA fragmentation in HepG2 cells, resulting from direct oxidation by the hydroxyl radical. *Biochem. J.* **317,** 13–16.

20. Su, J. H., Deng, G., and Cotman, C. W. (1997) Neuronal DNA damage precedes tangle formation and is associated with up-regulation of nitrotyrosine in Alzheimer's disease brain. *Brain Res.* **774,** 193–199.

21. Stadelmann, C., Bruck, W., Bancher, C., Jellinger, K., and Lassmann, H. (1998) Alzheimer disease: DNA fragmentation indicates increased neuronal vulnerability, but not apoptosis. *J. Neuropathol. Exp. Neurol.* **57,** 456–464.

22. Schulze-Osthoff, K., Walczak, H., Droge, W., and Krammer, P. H. (1994) Cell nucleus and DNA fragmentation are not required for apoptosis. *J. Cell Biol.* **127,** 15–20.

23. Zhang, G., Gurtu, V., Kain, S. R., and Yan, G. (1997) Early detection of apoptosis using a fluorescent conjugate of annexin V. *Biotechniques* **23,** 525–531.

24. Mesner, P. W. Jr. and Kaufmann, S. H. (1997) Methods utilized in the study of apoptosis. *Adv. Pharmacol.* **41:**57–87.

25. Gavrieli, Y., Sherman, Y., and Ben-Sasson, S. A. (1992) Identification of programmed cell death *in situ* via specific labeling of nuclear DNA fragmentation. *J. Cell Biol.* **119,** 493–501.

26. Tateyama, H., Tada, T., Hattori, H., Murase, T., Li, W.-X., and Eimoto, T. (1998) Effects of prefixation and fixation times on apoptosis detection by *in situ* end-labeling of fragmented DNA. *Arch. Pathol. Lab. Med.* **122,** 252–255.

27. Kockx, M. M., Muhring, J., Knaapen, M. W. M., and deMeyer, G. R. Y. (1998) RNA synthesis and splicing interferes with DNA *in situ* end labeling techniques used to detect apoptosis. *Am. J. Pathol.* **152,** 885–888.

28. Nakamura, M., Yagi, H., Ishii, T., et al. (1997) DNA fragmentation is not the primary event in glucocorticoid-induced thymocyte death *in vivo. Eur. J. Immunol.* **27,** 999–1004.

29. Willingham, M. C. (1999) Cytochemical methods for the detection of apoptosis. *J. Histochem. Cytochem.* **47,** 1101–1109.

30. Raina, A. K., Hochman, A., Zhu, X., et al. (2001) Abortive apoptosis in Alzheimer's disease. *Acta Neuropathol.* **101,** 305–310.

31. Stennicke, H. R., Jurgensmeier, J. M., Shin, H., et al. (1998) Pro-caspase-3 is a major physiologic target of caspase-8. *J. Biol. Chem.* **273,** 27,084–27,090.

32. Cohen, G. M. (1997) Caspases: the executioners of apoptosis. *Biochem. J.* **326,** 1–16.

33. Trucco, C., Oliver, F. J., de Murcia, G., and Menissier-de Murcia, J. (1998) DNA repair defect in poly(ADP-ribose) polymerase-deficient cell lines. *Nucleic Acids Res.* **26,** 2644–2649.

34. Roth, K. A., Kuan, C.-Y., Haydar, T. F., et al. (2000) Epistatic and independent functions of caspase-3 and Bcl-X$_L$ in developmental programmed cell death. *Proc. Natl. Acad. Sci. USA* **97,** 466–471.

35. Selznick, L. A., Holtzman, D. M., Han, B. H., et al. (1999) *In situ* immunodetection of neuronal caspase-3 activation in Alzheimer disease. *J. Neuropathol. Exp. Neurol.* **58,** 1020–1026.

36. Stadelmann, C., Deckwerth, T. L., Srinivasan, A., et al. (1999) Activation of caspase-3 in single neurons and autophagic granules of granulovacuolar

degeneration in Alzheimer's disease. Evidence for apoptotic cell death. *Am. J. Pathol.* **155**, 1459–1466.

37. Gervais, F. G., Xu, D., Robertson, G. S., et al. (1999) Involvement of caspases in proteolytic cleavage of Alzheimer's amyloid-β precursor protein and amyloidogenic Aβ peptide formation. *Cell* **97**, 395–406.

38. Pike, C. J. (1999) Estrogen modulates neuronal Bcl-xL expression and beta-amyloid-β-induced apoptosis: relevance to Alzheimer's disease. *J. Neurochem.* **72**, 1552–1563.

39. Srinivasula, S. M., Hegde, R., Saleh, A., et al. (2001) A conserved XIAP-interaction motif in caspase-9 and Smac/DIABLO regulates caspase activity and apoptosis. *Nature* **410**, 112–116.

40. Torp, R., Su, J. H., Deng, G., and Cotman, C. W. (1998) GADD45 is induced in Alzheimer's disease, and protects against apoptosis *in vitro. Neurobiol Disease* **5**, 245–252.

41. Martins, L. M., Kottke, T. J., Kaufmann, S. H., and Earnshaw, W. C. (1998) Phosphorylated forms of activated caspases are present in cytosol from HL-60 cells during etoposide-induced apoptosis. *Blood* **92**, 3042–3049.

42. Smith, M. A., Perry, G., Richey, P. L., et al. (1996) Oxidative damage in Alzheimer's. *Nature* **382**, 120–121.

43. Hampton, M. B., Fadeel, B., and Orrenius, S. (1998) Redox regulation of the caspases during apoptosis. *Ann. N.Y. Acad. Sci.* **854**, 328–335.

44. LeBlanc, A., Liu, H., Goodyer, C., Bergeron, C., and Hammond, J. (1999) Caspase-6 role in apoptosis of human neurons, amyloidogenesis, and Alzheimer's disease. *J. Biol. Chem.* **274**, 23,426–23,436.

45. Lassmann, H., Bancher, C., Breitschopf, H., et al. (1995) Cell death in Alzheimer's disease evaluated by DNA fragmentation *in situ. Acta Neuropathol.* **89**, 35–41.

46. Lucassen, P. J., Chung, W. C., Vermeulen, J. P., Van Lookeren Campagne, M., Van Dierendonck, J. H., and Swaab, D. F. (1995) Microwave-enhanced *in situ* end-labeling of fragmented DNA: parametric studies in relation to postmortem delay and fixation of rat and human brain. *J. Histochem. Cytochem.* **43**, 1163–1171.

47. Lucassen, P. J., Chung, W. C., Kamphorst, W., and Swaab, D. F. (1997) DNA damage distribution in the human brain as shown by *in situ* end labeling; area-specific differences in aging and Alzheimer disease in the absence of apoptotic morphology. *J. Neuropathol. Exp. Neurol.* **56**, 887–900.

48. Sheng, J. G., Mrak, R. E., and Griffin, W. S. (1998) Progressive neuronal DNA damage associated with neurofibrillary tangle formation in Alzheimer disease. *J. Neuropathol. Exp. Neurol.* **57**, 323–328.

49. Sheng, J. G., Zhou, X. Q., Mrak, R. E., and Griffin, W. S. (1998) Progressive neuronal injury associated with amyloid plaque formation in Alzheimer disease. *J. Neuropathol. Exp. Neurol.* **57**, 714–717.

50. Slivka, A., Mytilineou, C., and Cohen, G. (1987) Histochemical evaluation of glutathione in brain. *Brain Res.* **409**, 275–284.

51. Stadtman, E. R. (1992) Protein oxidation and aging. *Science* **257**, 1220–1224.

52. Stadtman, E. R. (1993) Oxidation of free amino acids and amino acid residues in proteins by radiolysis and by metal-catalyzed reactions. *Annu. Rev. Biochem.* **62**, 797–821.

53. Beckman, J. S. (1996) Oxidative damage and tyrosine nitration from peroxynitrite. *Chem. Res. Toxicol.* **9,** 836–844.
54. Alvarez, B., Rubbo, H., Kirk, M., Barnes, S., Freeman, B. A., and Radi, R. (1996) Peroxynitrite-dependent tryptophan nitration. *Chem. Res. Toxicol.* **9,** 390–396.
55. Monnier, V. M. and Cerami, A. (1981) Nonenzymatic browning in vivo: possible process for aging of long-lived proteins. *Science* **211,** 491–493.
56. Sayre, L. M., Sha, W., Xu, G., et al. (1996) Immunochemical evidence supporting 2-pentylpyrrole formation on proteins exposed to 4-hydroxy-2-nonenal. *Chem. Res. Toxicol.* **9,** 1194–1201.
57. Itakura K., Uchida, K., and Osawa, T. (1996) A novel fluorescent malondialdehyde-lysine adduct. *Chem. Phys. Lipids* **84,** 75–79.
58. Sakurai, T. and Tsuchiya, S. (1988) Superoxide production from nonenzymatically glycated protein. *FEBS Lett.* **236,** 406–410.
59. Beckman, J. S., Ye, Y. Z., Anderson, P. G., et al. (1994) Extensive nitration of protein tyrosines in human atherosclerosis detected by immunohistochemistry *Biol. Chem. Hoppe-Seyler* **375,** 81–88.
60. Sell, D. R. and Monnier, V. M. (1989) Structure elucidation of a senescence cross-link from human extracellular matrix. Implication of pentoses in the aging process. *J. Biol. Chem.* **264,** 21,597–21,602.
61. Xu, G. and Sayre, L. M. (1998) Structural characterization of a 4-hydroxy-2-alkenal-derived fluorophore that contributes to lipoperoxidation-dependent protein cross-linking in aging and degenerative disease. *Chem. Res. Toxicol.* **11,** 247–251.
62. Wells-Knecht, K. J., Brinkmann, E., and Baynes, J. W. (1995) Structural characterization of an imidazolium salt formed from glyoxal and Na-hippuryl-lysine. *J. Org. Chem.* **60,** 6246–6247.
63. Miyata, S. and Monnier, V. M. (1992) Immunohistochemical detection of advanced glycosylation end products in diabetic tissues using monoclonal antibody to pyrraline. *J. Clin. Invest.* **89,** 1102–1112.
64. Sayre, L. M., Arora, P. K., Iyer, R. S., and Salomon, R. G. (1993) Pyrrole formation from 4-hydroxynonenal and primary amines. *Chem. Res. Toxicol.* **6,** 19–22.
65. Reddy, S., Bichler, J., Wells-Knecht, K. J., Thorpe, S. R., and Baynes, J. W. (1995) N epsilon-(carboxymethyl) lysine is a dominant advanced glycation end product (AGE) antigen in tissue proteins. *Biochemistry* **34,** 10,872–10,878.
66. Shipanova, I. N., Glomb, M. A., and Nagaraj, R. H. (1997) Protein modification by methylglyoxal: chemical nature and synthetic mechanism of a major fluorescent adduct. *Arch. Biochem. Biophys.* **344,** 29–36.
67. Smith, M. A., Taneda, S., Richey, P. L., et al. (1994) Advanced Maillard reaction end products are associated with Alzheimer disease pathology. *Proc. Natl. Acad. Sci. USA* **91,** 5710–5714.
68. Yan, S.-D., Chen, X., Schmidt, A.-M., et al. (1994) Glycated tau protein in Alzheimer disease: a mechanism for induction of oxidant stress. *Proc. Natl. Acad. Sci. USA* **91,** 7787–7791.
69. Ledesma, M. D., Bonay, P., Colaco, C., and Avila, J. (1994) Analysis of microtubule-associated protein tau glycation in paired helical filaments. *J. Biol. Chem.* **269,** 21,614–21,619.
70. Perry, G. and Smith, M. A. (1993) Senile plaques and neurofibrillary tangles: what role do they play in Alzheimer's disease? *Clin. Neurosci.* **1:**199–203.

71. Trojanowski J. Q., Schmidt M. L., Shin, R.-W., Bramblett, G. T., Goedert, M., and Lee, V. M.-Y. (1993) PHF-τ (A68): from pathological marker to potential mediator of neuronal dysfunction and degeneration in Alzheimer's disease. *Clin. Neurosci.* **1,** 184–191.

72. Smith, M. A., Sayre, L. M., Anderson, V. E., et al. (1998) Cytochemical demonstration of oxidative damage in Alzheimer disease by immunochemical enhancement of the carbonyl reaction with 2,4-dinitrophenylhydrazine. *J. Histochem. Cytochem.* **46,** 731–735.

73. Montine, K. S., Kim, P. J., Olson, S. J., Markesbery, W. R., and Montine, T. J. (1997) 4-hydroxy-2-nonenal pyrrole adducts in human neurodegenerative disease. *J. Neuropathol. Exp. Neurol.* **56,** 866–871.

74. Smith, M. A., Kutty, R. K., Richey, P. L., et al. (1994) Heme oxygenase-1 is associated with the neurofibrillary pathology of Alzheimer's disease. *Am. J. Pathol.* **145,** 42–47.

75. Maines, M. D. (1988) Heme oxygenase: function, multiplicity, regulatory mechanisms, and clinical applications. *FASEB J.* **2,** 2557–2568.

76. Premkumar, D. R. D., Smith, M. A., Richey, P. L., et al. (1995) Induction of heme oxygenase-1 mRNA and protein in neocortex and cerebral vessels in Alzheimer's disease. *J. Neurochem.* **65,** 1399–1402.

77. Takeda, A., Perry, G., Abraham, N. G., et al. (2000) Overexpression of heme oxygenase in neuronal cells, the possible interaction with tau. *J. Biol. Chem.* **275,** 5395–5399.

78. Takeda, A., Smith, M. A., Avila, J., et al. (2000) In Alzheimer's disease, heme oxygenase is coincident with Alz50, an epitope of tau induced by 4-hydroxy-2-nonenal modification. *J. Neurochem.* **75,** 1234–1241.

79. Arrasate, M., Perez, M., Valpuesta, J. M., and Avila, J. (1997) Role of glycosaminoglycans in determining the helicity of paired helical filaments. *Am. J. Pathol.* **151,** 1115–1122.

80. Zhu, X., Rottkamp, C. A., Boux, H., Takeda, A., Perry, G., and Smith, M. A. (2000) Activation of p38 kinase links tau phosphorylation, oxidative stress and cell cycle-related events in Alzheimer disease. *J. Neuropathol. Exp. Neurol.* **59,** 880–888.

81. Zhu, X., Raina, A. K., Rottkamp, C. A., et al. (2001) Activation and redistribution of c-Jun N-terminal kinase/stress activated protein kinase in degenerating neurons in Alzheimer's disease. *J. Neurochem.* **76,** 435–441.

82. Raina, A. K., Zhu, X., Rottkamp, C. A., Monteiro, M., Takeda, A., and Smith, M. A. (2000) Cyclin' toward dementia: cell cycle abnormalities and abortive oncogenesis in Alzheimer disease. *J. Neurosci. Res.* **61,** 128–133.

83. Qian, Z. M. and Wang, Q. (1998) Expression of iron transport proteins and excessive iron accumulation in the brain in neurodegenerative disorders. *Brain Res. Rev.* **27,** 257–267.

84. Smith, M. A., Wehr, K., Harris, P. L. R., Siedlak, S. L., Connor, J. R., and Perry, G. (1998) Abnormal localization of iron regulatory protein in Alzheimer's disease. *Brain Res.* **788,** 232–236.

85. Lovell, M. A., Robertson, J. D., Teesdale, W. J., Campbell, J. L., and Markesbery, W. R. (1998) Copper, iron and zinc in Alzheimer's disease senile plaques. *J. Neurol. Sci.* **158,** 47–52.

86. Smith, M. A., Harris, P. L. R., Sayre, L. M., and Perry, G. (1997) Iron accumulation in Alzheimer disease is a source of redox-generated free radicals. *Proc. Natl. Acad. Sci. USA* **94,** 9866–9868.

87. Perez, M., Valpuesta, J. M., de Garcini, E. M., et al. (1998) Ferritin is associated with the aberrant tau filaments present in progressive supranuclear palsy. *Am. J. Pathol.* **152,** 1531–1539.
88. Rottkamp, C. A., Raina, A. K., Zhu, X., et al. (2001) Redox-active iron mediates amyloid-β toxicity. *Free Radical Biol. Med.* **30,** 447–450.
89. Sayre, L. M., Perry, G., Harris, P. L. R., Liu, Y., Schubert, K. A., and Smith, M. A. (2000) In situ oxidative catalysis by neurofibrillary tangles and senile plaques in Alzheimer's disease: a central role for bound transition metals. *J. Neurochem.* **74,** 270–279.

Index

From: *Neuromethods: Vol. 37: Apoptosis Techniques and Protocols, 2nd Ed.*
Edited by: A. C. LeBlanc © Humana Press Inc., Totowa, NJ